Industrial practices in weaving preparatory

Industrial practices in weaving preparatory

Dr Mukesh Kumar Singh

WOODHEAD PUBLISHING INDIA PVT LTD

New Delhi

Published by Woodhead Publishing India Pvt. Ltd.
Woodhead Publishing India Pvt. Ltd.,
303, Vardaan House, 7/28, Ansari Road,
Daryaganj, New Delhi - 110002, India
www.woodheadpublishingindia.com

First published 2014, Reprint 2017, 2018
© Woodhead Publishing India Pvt. Ltd., 2014

Woodhead Publishing India Pvt. Ltd. ISBN: 978-93-80308-29-6
Woodhead Publishing India Pvt. Ltd. e-ISBN: 978-93-80308-51-7

Typeset by Mind Box Solutions, New Delhi
Printed and bound by Replika Press Pvt. Ltd.

Contents

Weaving and weaving preparatory are part of the ancient era which includes various conventional and non-conventional practices. The purpose of this book is to present state of art with respect to the engineering and scientific aspects of yarn handling during winding, warping, sizing and drawing-in process.

This book approaches the subject of weaving preparatory from a scientific viewpoint being principally concerned with the relationship between weaving preparatory and weaving performance. Several recent developments in information technology, electronics and material science have indicated that there is currently strong worldwide interest in using the technology in weaving preparatory to make it more productive and precise such as winding, warping and sizing.

Research goals in these fields are driven by the demand of customer to produce zero-defect ideal fabrics. A welcome trend in recent years of scientists of other fields like biomaterials, material science is observed to develop new sizing materials to size a wide range of yarns. Varieties of alternate materials are explored to size a range of textile yarns made of different fibres. Biochemists have modified the starch molecules in different ways to make it suitable to size various textile yarns and desize it quickly without leaving any residual mark.

The general approach of this book is to treat the textile materials as highly anisotropic material 'in the jargon of material science'. In this way, the various process parameters in weaving preparatory processes play a vital role to produce a homogeneous product.

The coverage of chapters emphasizes a journey of conventional to most recent developments in continuous phases. The chapters in this book comprise an electric mix of applied, theoretical and engineering-oriented approaches to the problems of various weaving preparatory processes. This project aims at a description of the dynamic cause–effect relations between a machine and its product. Feeling the long-standing experience of the author, the main but not exclusive focus of this book is concentrated on the scientific aspect of yarn winding, emerging features of winding machines from conventional to era of nanotechnology, highly sensitive warping process and energy-consuming sizing process from machinery to process control viewpoint.

Lack of availability of book in the field of weaving preparatory encouraged me to write this book to provide maximum information from conventional to latest super high speed machines controlled by central computers. Efforts to modify starch, a common yarn sizing molecule, by various chemists and biochemists have also compelled me to present systematic knowledge of various sizing materials in the form of a book.

Readers of this book would find the conceptual aspects of winding, warping and sizing machines alongwith the importance of process parameters in order to produce a product of homogeneous characteristics. A variety of sizing processes like hot melt sizing, dye sizing, cold sizing, single-end sizing, etc., practiced in various weaving preparatory sections of various textile industries, are the part of content in this book. Different modifications made by various leading weaving preparatory machine manufacturers are also available to the readers. Assessment of sizing performance in the form of sized yarn is not covered systematically in easily available textile books. This book covers few modern techniques used in the assessment of sized yarn performance under actual weaving conditions. Various technocrats working in this field, academecians and students at both under- and postgraduate level are the target beneficiaries of this book. It took almost 2 years to complete this book.

I feel these goals have been truly achieved. This book is coming in the market when my institute is going to celebrate 100 years of its foundation.

I am really thankful to my family to support me to complete this book restlessly.

Dr Mukesh Kumar Singh
UP Textile Technology Institute
(Formerly GCTI), Kanpur, India

1

Yarn winding for warp and weft

1.1   Introduction and objectives

Yarn produced either by spinning, staple, and fibers or by extruding filaments from polymeric materials needs to be wound on a suitable package. The formation of such packages comes under the area of yarn manufacturing. Here, the discussion is focused on the suitability of packages rewinding. Ring bobbins (Fig. 1.1) formed during spinning are converted into packages known as cone (Fig. 1.2). These are of desirable shape, build, and quality that would be compatible with the subsequent processes (Fig. 1.3).

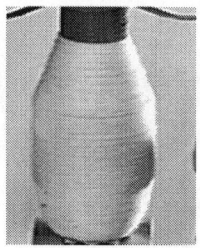      

Fig. 1.1 Ring frame bobbin   **Fig. 1.2** Cone   **Fig. 1.3** Double flange bobbin

A typical ring bobbin of cotton yarn can be of 180–360 mm in length and up to 72 mm in diameter, containing 2000–4000 m of yarn weighing about 80–120 g. On the other hand, a cone can be of 90–150 mm height with a base diameter of nearly 300 mm and cone angle varying from 4°20' to as high as 11° carrying about 2–3 kg of yarn of 50–100 km length. The objectives of winding can be briefly stated as:

- Removing objectionable faults
- Building a package of suitable dimensions
- Keeping safe the yarn surface during winding process

Additionally, one would also like to ensure that the surface of the yarn is not damaged in any way during the process of winding (e.g. increasing hairiness), and if possible, can be improved (e.g. through waxing). The winding of yarn must be done in such a way as to permit unwinding in the following processes with a minimum of difficulty at the required speed. Moreover, the package shape, size, and build must be the most suitable for the particular end use.

1.2   Classification of winding machines

Winding machines can be classified in different ways and few of them are discussed below:

1.2.1   On the basis of winding

On the basis of winding, winding machines can be classified into two groups:
1. Precision winding
2. Non-precision winding

1.2.2   On the basis of package type

On the basis of package type, winding machines can be classified into at least five groups as given below:
1. Cope winding machines
2. Cone winding machines
3. Cheese winding machines
4. Flanged bobbin winding machines
5. Pirn winding machines

1.2.3   On the basis of package drive

1. Direct drive winding machines
2. Indirect drive winding machines

1.2.4   On the basis of yarn type

(a) Warp yarn winding machines
   (i) Upright spindle winding machine
   (ii) Drum winding machine
(b) Weft yarn winding machines
   (i) Pirn winding machines
      (a) Ordinary pirn winding machines
      (b) Automatic pirn winding machines
   (ii) Cope winding machines

1.3   Package build

Packages may be broadly categorized into two groups, namely (1) flanged and (2) flangeless. A flanged package (Fig. 1.3) is equipped with two discs mounted on the package tube. These discs are meant to provide lateral support to the yarn wound in the intervening space. The flangeless package (Fig. 1.2) does not exhibit any such support. The build of the package, i.e. the exact manner in which yarn is wound around and along the tube, is therefore very critical in case of the flangeless packages, as the yarn mass has to be self-supporting. In the following, therefore, the build of the package would be considered from the point of view of flangeless packages only. Flangeless packages permit yarn unwinding at very high speed through over end withdrawal.

Flangeless packages can in turn be grouped into two classes namely:

(1) Parallel-wound
(2) Cross-wound

Parallel-wound flangeless packages are invariably tapered at the two ends. As a result, a coil of yarn at the boundary layer of any diameter is supported against lateral slippage by yarn coils lying just ahead in the succeeding layers. In the process, however, some precious space is lost, leading to a lowering in the yarn content of the package.

The parallel-wound packages may be built in three ways namely, cope build (Figs. 1.4 and 1.5), roving build (Fig. 1.6), and combination build (Fig. 1.7).

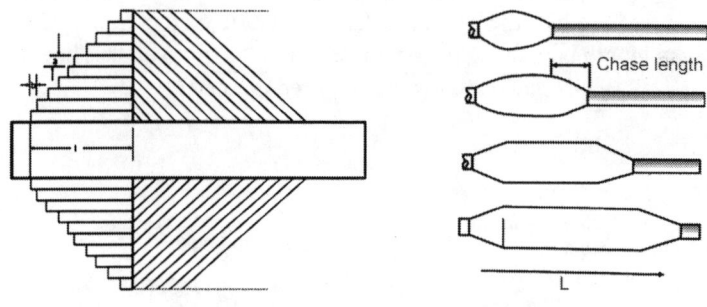

Fig. 1.4 Cope build   **Fig. 1.5** Cope build

Cope-built packages, such as a pirn, are constructed from one end of the tube with a very short yarn traverse (Fig. 1.4). After few layers are wound over a short length of the tube and a specific diameter is built up, the yarn traversing mechanism is shifted by a definite amount and winding continues, partly on a stretch of bare surface of the tube and partly on the previously constructed layers.

By programming the diameter sensing system, it is possible to control the taper angle at the beginning of the cope. Because of the taper at the beginning, each layer of yarn wound subsequently in the body of the tube has to exhibit a slope inclined in the opposite direction. This slope can be varied throughout the construction of the package or may be kept constant. The length measured along this slope projected on the axis of package is termed the chase length. The cope-built package is thus built up by continuously adding small bits of elements from the base towards the nose. The shape of the typical hump of such a package (Fig. 1.5) can take different shapes depending on the contour of the bare tube.

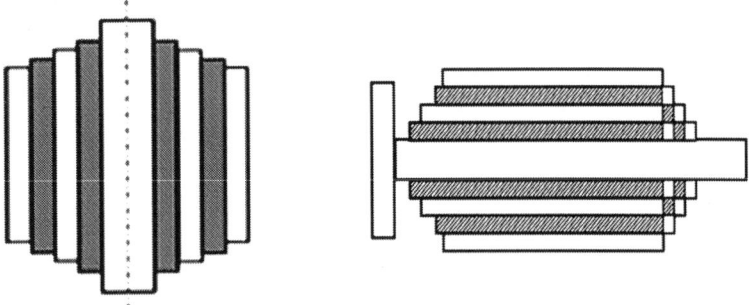

Fig. 1.6 Roving built package   **Fig. 1.7** Combination built package

In roving built packages (Fig. 1.6), the yarn is traversed at the beginning along the entire length of the bare tube. Once a certain diameter is built up, the total traverse length is reduced by a definite amount at both ends. This process is continued till the package attains the required diameter.

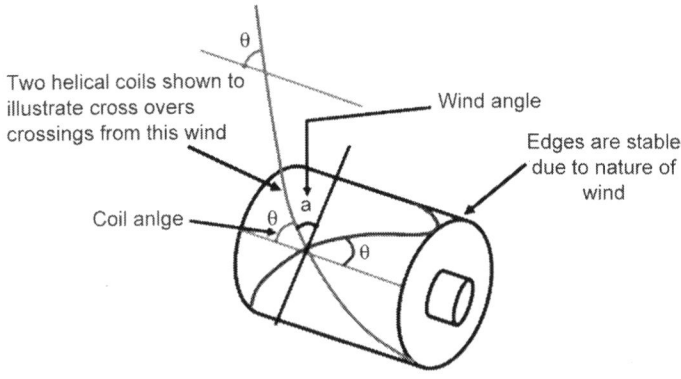

Fig. 1.8 Cross wound package

The combination built package is a fusion of the cope and roving build. In this type of built (Fig. 1.7), the traverse at the beginning is much greater than that in the case of the cope build (Fig. 1.4); although, it is not extending the whole length of the tube. Thus each yarn layer covers a long length of the tube so much so that even while unwinding from the last few layers, the nose of the residual package is quite close to the tip of the tube (Fig. 1.11).

Commercially produced cross-wound packages are of two types, namely cheese and cone. All cross-wound packages have very sharp walls at the package base and nose (Fig. 1.8). These walls may either be normal to the plane of the package axis or may even be curved. Cross-wound packages exhibit very large angle of wind and very low number of coils per traverse. The length of traverse and the traverse speed are usually constant throughout the package build; although in pineapple cone, the traverse is slightly reduced as the diameter increases. A cheese is built on a hollow cylindrical core, whereas a cone is constructed on a frustum. The taper of the cone or the conicity may remain constant throughout the package or may gradually rise, as in packages with accelerated taper.

1.4   Few important terms of winding

1.4.1   Wind angle or winding angle

It is the angle at which the yarns cross each other on a wound package. Few textile scientists defined it in different ways. According to them, the wind angle is half of the angle at which the yarns cross each other on a wound package as shown in Fig. 1.8. Smaller the crossing angle, denser will be the package.

1.4.2   Winding ratio or wind

It is the number of coils of yarn per traverse of yarn from one extreme end to another extreme end of the package. More the number of coils per revolution more is the density. Increasing the distance between the threads of a layer decreases the density but increases the package stability.

1.4.3   Package stability

Package stability comes from the tendency of wound yarn on the package. If the yarn layers hold properly on the package, the package will remain stable in subsequent processes. For example, in a typical case of precision winding machines, the wind angle reduces with increase in package diameter; consequently, the packages become a little bit harder from outside layer and package stability reduces.

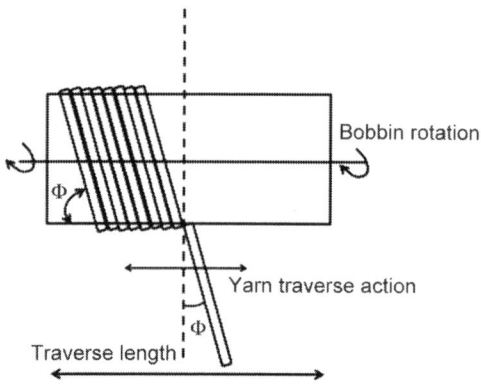

Fig. 1.9 Basic winding actions

The following factors also affect the package stability:
- The wind angle can usually be varied between 30° and 50°. The larger the wind angle, lower the density and greater the softness of the package.
- The wind angle mainly affects the package density and the unwinding performance of the package.
- The wind angle 40° is suitable for cones.
- The wind angle varies from 33 to 36° for cylindrical packages.
- The wind angle 40–45° is appropriate for packages dedicated for package dyeing.
- In case of latest high temperature high pressure (HTHP) dyeing machines, dyeing can be carried out even at 30–35° wind angle packages.
- The package density depends not only upon wind angle but also on winding tension. The winding tension depends upon:
    - The contact pressure of the package on the winding drum
    - The yarn fineness (finer yarns give higher density)

Table 1.1 Common taper angles for random wound cones.

Cone taper semi-vertical angle	End uses
3°30'	General purposes
4°2'	Cones for package dyeing
5°5'	Weft knitting: final diameter cone taper may increase to 10°
9°1'	Weft knitting: final diameter cone taper may increase to 14–18°

1.4.4   Winding speed

It depends upon the following factors:
- Count
- Type of yarn, (type of fiber, average strength, and minimum strength)
- Type and characteristics of yarn supply bobbin
- Package taper
- Final use of package

The best winding speed is the speed which allows the highest level of production possible for a given type of yarn and type of package, and with no damage to the yarn (abrasion and breaks due to excessive tension).

1.4.5   Winding production

It depends upon the following factors:
- Winding speed
- Time required by the machine to carry out one splicing operation
- Bobbin length per bobbin (both bobbin weight and TPI to be considered, because TPI will affect the bobbin length). This decides the number of bobbin changes.
- The number of faults in the yarn and the clearer settings decides the clearer cuts
- Count
- The number of doffs depends upon the doff weight. Higher the doff weight, lower the number of doffs.
- The time taken for each doff, either by the doffer or by an operator
- Down time due to red light. It depends upon: number of red lights, number of repeaters, settings for red lights, clearer settings like off count channel, and cluster setting which will result in red lights and others.
- Bobbin rejections depend on weak yarn, wrong gaiting, double gaiting, and bobbin characteristics, etc.

1.5   Unwinding behavior

During over end unwinding, fluctuation in unwinding tension (Fig. 1.10) is primarily caused by the phenomenon of balloon formation. Accordingly, the unwinding tension is directly proportional to the square of the balloon height (Fig. 1.11) and inversely proportional to the square of the unwinding radius.

Insert some relationship

T directly proportional to h^2

T indirectly proportional to $1/r^2$

$T = kh^2/r^2$

The major causes of unwinding tension fluctuations in various types of packages are as under (Foster 1959):

- In case of flangeless parallel wound packages, it can be inferred that the short-term fluctuation in unwinding tension, i.e. within a traverse, would primarily be caused by the change in radius of unwinding point between the nose and the base of a cope-built package.
- In case of roving built package, the change in balloon height would be the source of tension fluctuations. The roving built package would also exhibit a continuous rise in unwinding tension, caused by the gradual fall in unwinding radius. However, the corresponding rise in tension with roving built would be much less than in the cope-built packages. Thus the roving built package is advantageous from the point of view of long-term tension increase, whereas the cope built one is better in terms of the short term.
- The unwinding behavior of pirn is very complex in nature when yarn unwinding takes place from its base, i.e. near to exhaustion. The short-term fluctuation in tension in pirn would be much lower compared to that in a tube of filament yarn.
- Evidently a hybrid built package provides a good compromise, if designed properly. The combination built package therefore provides a much steadier average value of unwinding tension.
- On one hand the balloon height has by then increased to a very high value, leading to a nearly exponential rise in tension, while on the other hand there is a severe problem of licking of the tube by the balloon, causing an unwarranted addition in tension.
- The cone provides the better solution in so far as unwinding is concerned. In a cone, the short-term tension variation is caused by continuous change in both the unwinding radius, as well as in the balloon height; whereas in a cheese, the change in balloon height only should be the cause. However, there is remarkable licking by yarn around the nose of a cheese leading to sharp peaks in unwinding tension. Hence the taper in conocity proves beneficial in suppressing the short-term tension variation.
- The long-term tension fluctuation in both the cheese as well as the cone is caused by a gradual reduction in the package diameter. Hence, it is very difficult to select the best between the cone and the cheese.
- The cross wound packages have very large angles of wind. During reversal of a traverse, there is an abrupt change in wind angle as well. Thus the wind angle quickly drops to a value of zero and then sharply increases in magnitude but in an opposite sense over a distance of few

millimeters only near the two package walls. This also has to exhibit the effect of causing a sudden and large periodic tension variation in the cross wound package.

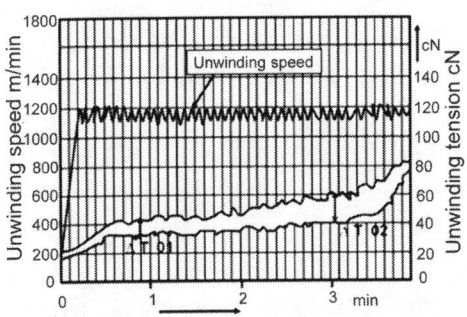

Fig. 1.10 Fluctuation in unwinding tension

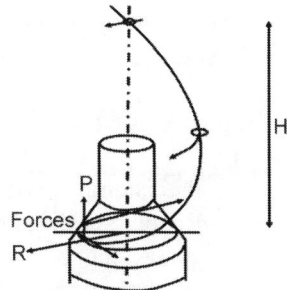

Fig. 1.11 Unwinding

- The unwinding tension is independent of the winding tension. Moreover at least in case of staple yarns and up to a certain extent in case of continuous filament yarns, greater winding tension reduces sloughing off. So in case of staple yarns, maximum winding tension should be used, however, taking care not to influence the yarn quality adversely.
- The initial tension when the yarn is pulled from a shuttle with a sudden jerk is not higher than that while unwinding continuously.
- Maximum tension occurs while unwinding at the cone apex.

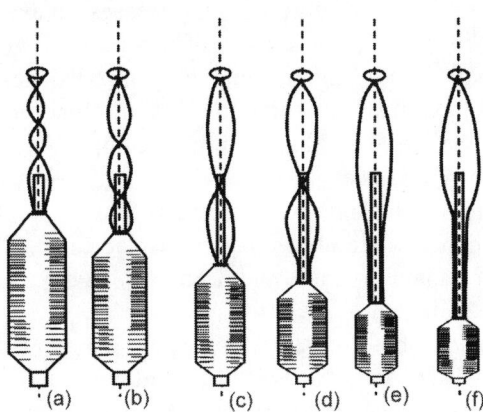

Fig. 1.12 Unwinding progress

1.6   Classification of winding process

A yarn package can be generated by rotating the same about its own axis and imparting a lateral to-and-fro motion to the yarn at the point of winding. The traversing motion of yarn affects reciprocation of yarn between the two extreme edges, along the length of the package. Continuous rotation of the package as well as traversing of the yarn may be achieved in a number of ways.

Principally, the winding strategy can be classified into two classes as given below:

- Precision winding
- Random or drum winding

A package may be rotated by either of two methods, namely

- By pressing the package onto a rotating drum, commonly termed as surface drive or drum winding (Fig. 1.13)
- By mounting the package directly on a rotating spindle, commonly termed as spindle drive or precision winding (Fig. 1.13)

1.6.1   Precision winding

In precision winding, the package is mounted on a spindle as shown in Fig. 1.13, which is driven with the help of an individual motor or machine shaft and yarn is displaced with the help of a guide or propeller blades (yarn laying system with counter rotating blades). There is zero uncontrolled displacement of the yarn; therefore, it is called precision winding. These machines are equipped with electronic or digital yarn tension control devices which vary the tension as the diameter of the package increases, so that the layer to layer density remains constant. Each drum is driven through an individual motor which in combination with a micro computer controlled meter counter, results into much more flexibility of operations. The machine productivity is high due to high running speed and lesser number of operators is required to run the same number of machines.

Major features of precision winding are as follows:

- The yarn guides are used for yarn displacement. The winding angle does not remain the same and decreases slightly with the increase in package diameter.
- There is no ribbon formation.
- Hard edges formation does not occur at the ends.
- Layer to layer density is very uniform throughout the package.

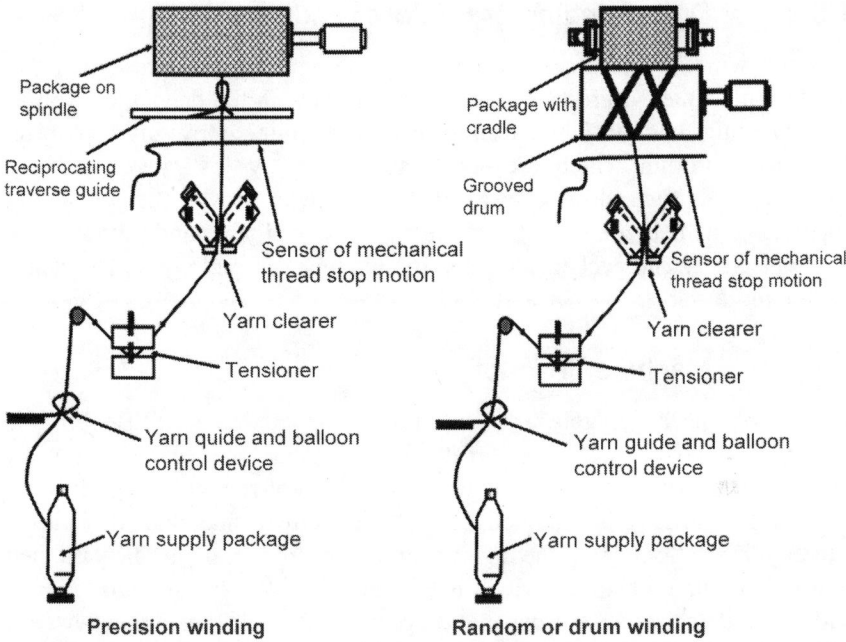

Precision winding   Random or drum winding

Fig. 1.13 Precision and random winding

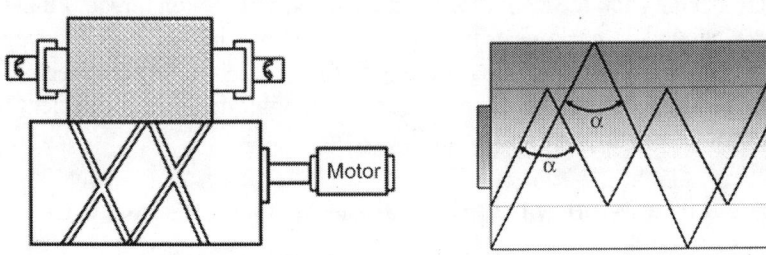

Fig. 1.14 Package rotation by   **Fig. 1.15** Yarn winding mechanism
surface contact   on drum-driven package

- The machine speed is high.
- Cost of the machine is medium to high.
- The number of coils per turn of the package remains same as the diameter of the package increases, because the number of displacement remains constant per rotation of the package.

1.6.2   Drum winding or random winding

Random winding machines are conventional winding machines. In drum winding machines, a grooved drum performs as both the driving as well as the traversing function. The package gets drive indirectly by surface contact with grooved drum, which is called friction drive. By this type of winding, the package is formed by a single thread which is laid on the package at appreciable helix angle so that the layers cross one another and give stability to the package. The packages formed by this type of winding are less dense but are more stable.

1.6.3   Machine description

The supply package, generally the ring frame bobbin (R/F Bobbin), is mounted on bobbin holder automatically or manually in creel section of machine, depending on the machine generation. A conventional self-adjusting balloon control ring is mounted over R/F bobbin to maintain the balloon height uniform throughout the unwinding of yarn from bobbin. The yarn then passes through a tensioner, which is essential to check the strength of yarn and to maintain the package density to a prescribed level. Tensioning device is followed by mechanical or electronic type (optical or capacitance based) yarn clearing system to prevent the passage of thick places ahead on final package. Then the yarn passes through a yarn sensing system, attached with thread stop motion to lift the package from drum surface in case of broken thread. Finally, the yarn is wound on package as per winding scroll of winding drum. Package is mounted over grooved winding drum with the help of cradle equipped with suitable bearing to achieve a smooth trouble- and vibration-free movement of package for a long time, as shown in Fig. 1.14.

1.6.4   Precision winding machine

Precision winding machine is able to laid successive coils of yarn close together in a parallel or near parallel manner. By this process, it is possible to produce very dense package with maximum amount of yarn stored in a given volume. Generally, precision winding package requires support from both edges of the package. In case of precision winding machine, package is directly mounted on driving spindle, and a separate yarn traversing system gives traverse to the yarn to cover the full width of package.

Yarn supply takes place from a ring bobbin or hank. A balloon control ring is provided to control the balloon height from actual point of unwinding. During yarn unwinding, the point of unwinding changes from top to bottom and then bottom to top alternately and consequently yarn tension also changes, which is a major hindrance in producing an ideally uniform density package. In case of modern precision winding machine, feature of changing height of balloon control ring according to the change in yarn unwinding point is also introduced to minimize the tension fluctuation as much as possible. After passing the balloon control ring, the yarn passes through tensioning device; here, further reduction in tension variation as well as addition of some tension on running yarn takes place. After that, yarn passes through yarn clearer device where yarn diameter scanning takes place. Yarn clearer may either be capacitance or optical based. Objectionable thick portions of yarn are eliminated here. Then yarn passes through feeler of thread stop motion which is required to give a signal to the operator in case of thread breakage due to any reason. Finally, the yarn passes through the yarn guide mounted over reciprocating traverse bar.

Main features:
- Package is wound with a reciprocating traverse.
- Patterning and rubbing cause damage of packages.
- Package contains more yarn.
- Package is less stable.
- The package is hard and compact.
- The package is dense.
- Rate of unwinding of package is low and the process of unwinding is hard.
- The unwound coil is arranged in a parallel or near parallel manner.

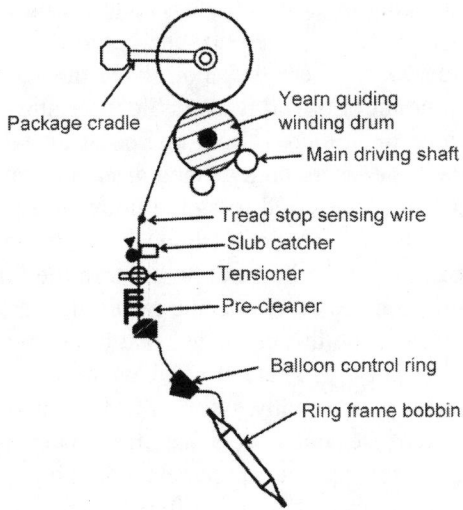

Fig. 1.16 Passage of material on drum winding machine

Fig. 1.17 Grooved winding drum

Generally, in cone and cheese winding, the yarn end is taken over from supply package and yarn tension is managed by tensioner directly, as shown in Fig. 1.16. In case of drum winding machines which work on constant rate of delivery, the yarn is supplied at a constant speed rather than a constant tension, and the tension is dependent on the ratio between the rates of yarn take-up and yarn delivery. The speed ratio is the ratio between the surface speed of the driving and delivery rollers. The winding ratio is the ratio between the traversing speed and the surface speed of the driving roller. On a cone winding machine, the rotational speed has a unique value for the whole cone but the surface speed varies, being highest at the base and lowest at the nose of the cone. Thus, at any one time, there is only one point of contact between the driving roller and the cone, where they both have the same surface speed. Thus, a very small amount of bound is available to be slip over nearby cone surface, and at any one time, there is only one point of contact between the driving roller and the cone where both driving drum and cone have the same surface speed. This point is referred to as the point of drive (POD). Under stable running conditions, the location of POD is usually found between the base and the nose of the cone or beyond the base. In the former case, it is a real point where, between the POD and the nose, the surface speed of the cone is lower than that of the driving roller, and between the POD and the base it is higher. POD is a hypothetical point which is used to understand the winding mechanism conceptually. As the POD is shifted away from the base, the surface speed of the cone lowers than the driving drum. Seed (1977) studied the influence of some process parameters on the position of the POD in order to relate it to the production of different winding machines; while

Paul (1982) studied about variations in the length of yarn being wound onto the cone during the winding process. Paul's study was aimed to reduce the amount of residual yarn on cones during warping process without considering the effect of yarn tension. Paul concluded that the winding speed (i.e., the position of the POD) was independent not only of the yarn tension, but also of the coefficient friction between driving roller and package.

The main characteristics of drum winding are given below:

1. *Winding angle* – Since grooved drum is used for winding, the winding angle remains constant through the process.
2. *Wind* – The number of coils per turn of the package decreases as the diameter of the package increases; consequently the package remains softer at the outer layers.
3. *Ribbon formation* – Ribbon formation may occur at certain points within the package, when diameter of package reaches to a certain ratio to drum diameter.
4. The package stability is good.
5. Layer to layer density is not very uniform throughout the package.
6. Hard edges formation may occur at the yarn, turning at the extreme ends of the drum.
7. Machine speed – The machine speed cannot be very high, because it is very difficult to maintain good density profile at high speeds on random winding machines.
8. Principally, the cost of machine is low.
    - Cross winding technique is used
    - The package density is low.
    - Minimum number of yarn is wound.
    - The package formed is soft and less compact.
    - The stability is high.
    - Flanges are not required.
    - The rate of unwinding is high and the process is easy.
    - The packages formed have low density.

1.6.5   Features of drum winding machine

- *Traverse of the package* – While winding particularly on the solid perforated cheeses, the yarn layers must cover only the perforated part of cheeses on both sides equally, because winding on blind areas will prevent the flow of dye liquor during package dyeing.
- *Ribbon breaking or anti-patterning device* – The machine must be equipped with a good and effective anti-patterning device to

minimize the possibility of ribbon formation. The ribbons formed during winding not only result into poor dyeing due to liquor flow obstruction, but also cause problems in unwinding by creating high tension zones and resulting into thread breakage, high hairiness due to increase friction, and wastages. However, problem of ribbon formation could not be eliminated completely till date.

- *Weight release mechanism* – Since the weight of package increases with the increase in the diameter of the package which increases the cradle pressure on the drum and affects the density of the packages. Therefore, the machine must be equipped with a proper weight-release mechanism to balance out the increased package weight (new machines are equipped with a programmed cradle weight regulation system).

- *Hard edges formation* – As per basic nature of drum winding with the groove drum, the yarn stays for a relatively longer time at the edges and creates hard edges. The machine must be equipped with a proper arrangement in which the cradle or the bobbin holder slides slightly (by few mm) on the drum through a lateral movement device which reduces the chances of hard edges formation by widening the edge area.

- *Length measuring device* – The machine must be equipped with a good length measuring system to minimize the package weight variation. The uniform package weight helps in reducing the weight variation in dyeing and also reduces the chances of excess dyeing and short falls.

- *Tension management* – A modern yarn winding machine must be equipped with an effective tension management system. It is very important to achieve uniform package density.

- *Balloon breakers* – The machine must be fitted with adequate balloon-breaking devices, which provide more flexibility to use different sizes of feed packages and also reduce the breakage rate. Balloon-breaking devices are helpful to maintain yarn tension throughout the unwinding of yarn from nose to base of R/F bobbin or other supply packages.

- Winding speed – The winding speed is also a constraint in random winders, but now machines are available with individual drum drives which can run at speed, and it is also possible to make soft packages on automatic winders which can run up to 1400 m/min. The cost, flexibility of running different counts, etc., and quality parameters made this drum winding machine a popular one in textile industry.

With the grooved drum system, the surface speed of the drum and the traverse speed are kept constant. A continuous helical groove (i.e., interconnected clockwise and counterclockwise helical grooves) around the drum circumference guides the yarn along the traverse length as the yarn is wound onto the bobbin.

A continuous helix has points of crossover of the clockwise and counterclockwise helices. To retain the yarn in the correct groove during its traverse, particularly at the intersections, one groove is made deeper than the other, and the shallower groove is slightly angled. For both types of traverse, we can refer to a drum constant, k, as the number of turns corresponding to a double traverse of the yarn on the bobbin. This is similar to the wind.

$$k = \frac{N_d}{N_t}$$

N_d is the drum rotational speed, N_t is traverse cycle per minute or traverse frequency.

A wing cam traverse (traverse for drum winding machine by means of a wing cam through a traverse bar and traverse guide) provides one double traverse (i.e., a traverse cycle) for every turn of the camshaft. There are therefore N_t traverse cycles per minute (i.e., the traverse frequency). The value of constant k is generally not a whole digit, but it can be. The value of k for a grooved drum is remained fixed and can be inspected easily.

The value of k remains twice of number of crossings of the spirals (i.e., wind). For example, the value of k for a drum which makes 2.5 spirals of thread when moving from one extreme end to opposite extreme end will be $2 \times 2.5 = 5$

Transverse ratio (TR)

$$TR = k\frac{D_d}{D_b}$$

Where D_d and D_b are the diameters of drum and bobbin, respectively.

The equation number is referred to the cylindrical packages. In case of cone, the mean cone diameter (d_m) is considered as effective package diameter. Hence for cone, the transverse ratio (TR) will be:

$$Tr = k\frac{D_d}{d_m}$$

1.6.6   Winding parameters

The bobbin length over which yarn is wound is termed the traverse length. The number of wraps (or coils) of yarn wound within a traverse length is called the wind; and the traverse ratio (TR) equals twice the wind, which is equal to the number of bobbin rotations in one traverse cycle. For the sake of simplicity, we first consider parallel wound packages and discuss cones later.

The angle θ between the inclined yarn (the yarn laid) on the package, and a plane perpendicular to the bobbin axis is called the wind angle and can be calculated according to:

$$\tan \theta = \frac{V_{ts}}{2\pi r N_b}$$

Where:

V_{ts} = the traverse speed (m/min), assumed to be constant even at the points of reversal

r = radius (m) of the layer being wound

N_b = bobbin rotational speed (rpm)

It is often found that the greater the wind angle, the more stable the package. The maximum limit to the wind angle is the value that, if exceeded, allows the yarn, on reaching the end of a layer during traverse reversals, to slip over the end of the layer beneath.

The coil angle φ is the angle between the direction of the yarn on the package and the direction of the traverse length. Therefore,

$$\phi + \theta = 90°$$

Throughout the remaining sections, the coil angle will be used in preference to the wind angle. The winding speed of the yarn is the resultant speed of the bobbin surface and traverse speed. It can be calculated from:

$$V_{wy} = \sqrt{V_{bs}^2 + V_{ts}^2}$$

Where:

V_{bs} = bobbin surface speed

V_{ts} = traverse speed

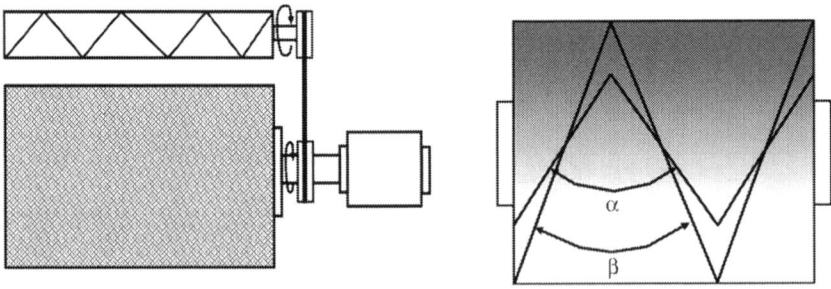

Fig. 1.18 Spindle drive cross winding

Table 1.2 Comparison between drum and precision winding.

Drum winding	Precision winding
The drum rotates at constant rpm; hence, its surface speed is constant.	The package rotates by spindle at constant rpm; hence its surface speed increases as package diameter increases.
Assuming no slippage between the drum and the package, it can be inferred that the package would have the same surface speed as that of the drum.	The role of slippage is nullified. Package has same rpm as that of spindle.
The frictional force μ.N is the prime mover of the package, where μ is the coefficient of friction between the surface of the package and the drum, while N is the normal force acting on the drum-package contact line (Fig. 1.21). Hence, the coefficient of friction between the drum surface and the yarn should be high enough to enable such a system to function; alternately, the normal pressure should be very high.	The package gets its motion directly from spindle; hence the role of friction and force is nullified.
Yarns of smooth and delicate surface should not be wound on surface driven machines, as the unavoidable slippage between the drum and the layers of yarn would not only damage the yarn surface but also disturb the winding process leading to defective packages.	Yarns of smooth and delicate surface can be wound safely on precision winding because the possibility of slippage and friction is zero in this process.
The rpm of a surface-driven package would keep on diminishing in a hyperbolic fashion with increase in package diameter.	Spindle-driven packages rotate at constant rpm. Obviously, the yarn surface character does not play any role in the building up of package.
When package diameter increases, the surface speed of package remains constant, hence yarn tension will also remain constant.	The package diameter keeps on increasing, the surface speed would also rise leading to a rise in winding speed and hence in yarn tension.
There is no need to restrict the package diameter or gradual reduction in package rpm with increase in the package diameter. The bigger size packages can be produced on this type of winding machine.	The problem of increasing yarn tension can be circumvented by either restricting the maximum package diameter to a moderately low value or by designing a closed loop system that would ensure a smooth and gradual reduction in package rpm with increase in the package diameter. The latter solution is expensive, whereas the former may prove uneconomical.

Contd...

Contd...

Drum winding	Precision winding
Since grooved drum is used for winding, the wind angle remains constant through the process.	As package rpm and traverse ratio remains constant, wind angle diminishes as the package diameter increases.
The number of coils per turn of the package decreases as the diameter of the package increases as shown in Fig 1.15; consequently the package remains softer at the outer layers.	The number of coils per turn of the package remains constant as the diameter of the package increases as shown in Fig 1.18; consequently the package remains harder at the outer layers.
Ribbon formation may occur at certain points within the package when diameter of package reaches to a certain ratio to drum diameter.	Possibility of ribbon formation is negligible because traverse guide movement is controlled separately.
The number of crossings between yarns of two adjacent layers would keep on diminishing.	The number of crossings between yarns of two adjacent layers would remain constant.

1.7   Hybrid winding (step precision winding)

After eliminating the technological problems of both the random and precision winding by recent advancement in precision winding by step precision technology, a new winding concept is developed and named as step precision winding or hybrid winding or digicone winding (Fig. 1.19).

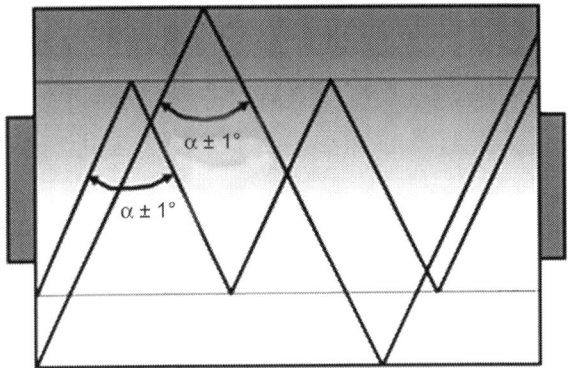

Fig. 1.19 Hybrid wind strategy on package

The major features of hybrid winding system are as follows:

The winding angle remains almost constant throughout the winding from empty to full package.

1. The package stability is also good.
2. Package remains constant with change in package diameter.
3. The probability of ribbon formation is close to zero.
4. The unwinding performance from hybrid package is very good.
5. The cost of this type of winding machine is comparatively high.

1.8 Yarn traverse system

To wind the yarn on full working length of winding package uniformly, yarn traversing mechanism is required.

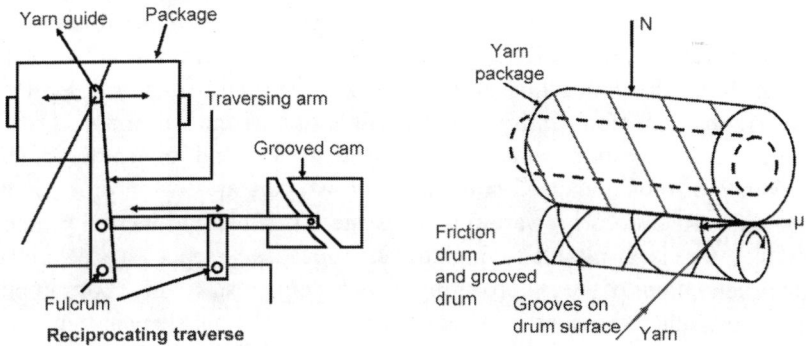

Fig. 1.20 Reciprocating traverse   Fig. 1.21 Rotary traverse

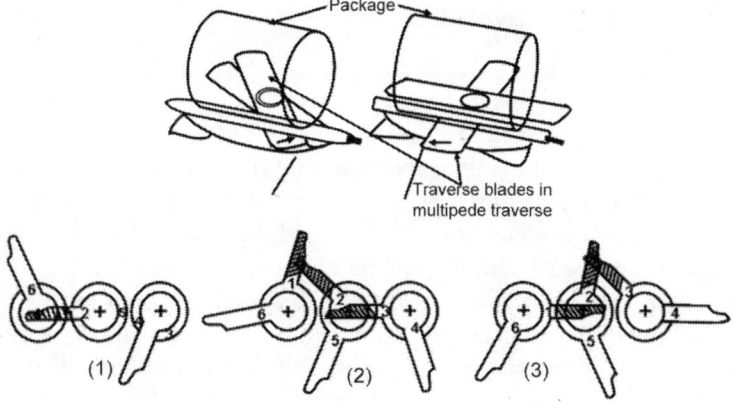

Fig. 1.22 Multipede traverse

Principally, yarn traverse mechanisms are classified into three major classes:

- *Reciprocating traverse:* Reciprocating a yarn guide along the length of the package, commonly called as reciprocating traverse (Fig. 1.20).
- Guiding the yarn by a complex of helical grooves designed on the surface of a rotating drum, commonly called as rotary traverse (Fig. 1.21).
- Guiding the yarn by a series of rotating blades called multipede traverse (Fig. 1.22).

1.8.1   Reciprocating traverse

The reciprocating traverse system essentially consists of a grooved cam to convert circular motion to reciprocating movement. Grooved cam is connected with reciprocating horizontal shaft, with the help of a bowl and fulcrum. The horizontal shaft is connected with a bottom fulcrum vertical shaft as shown in Fig. 1.20. The top of this vertical shaft is connected with a thread guide to traverse the yarn along entire winding length of package uniformly.

The reciprocating yarn traverse system is found on rotor spinning machines or on slow-speed non-automatic winders as also on pirn winding machines. Reciprocating yarn traverse system is subject to inertial problems and therefore is unsuitable for high traverse speed. The repeated accelerations and decelerations of the reciprocating mass not only result in high consumption of energy, but also cause wear and tear of the driving elements. A typical traverse speed on a 12,000 rpm pirn-winding machine would be in the range of 40–50 m/min.

1.8.2   Rotary traverse

The rotary traverse system essentially consists of a grooved winding drum to traverse the yarn from one end to opposite end of the package. The winding package rests over grooved drum, as shown in Fig. 1.21. Numbers of grooved drums are mounted on winding machine shaft at definite intervals for each winding head on conventional type drum winding machine. The depth and width of groove on winding drum surface are designed in such a way to ensure the uniform traverse of yarn throughout the process. Generally, cross wound packages are formed by rotary traverse. The major drawback of this system is the continuous abrasion of yarn against drum surface. This system avoids the possibility of jerks and inertia effect. The speed of rotation of the drum determines the traverse speed, and its maximum value would be governed by factors such as the design of the groove, the material of the drum, the rigidity of the driving elements, etc. A typical traverse speed on a 4,000 rpm drum of 1.5 scrolls producing a 150 mm tall package would be 400 m/min.

1.8.3   Multipede traverse

Multipede traverse system combines the principles of reciprocating and rotary traverse, leading to the possibility of a much higher traverse speed. In such a system, rotating blades (Fig. 1.22), which are synchronized properly, guide the yarn to-and-fro. Very little mass is involved here and there, and there is no change in direction of rotation either. Such systems are usually employed in very high speed winders needed on filament spinning machines.

SSM-PSM-51 precision winder has used two counter rotating blades to traverse the yarn. Blades rotate at constant speed and traverse the yarn alternately to wind it on package uniformly. The required density profile of the package is achieved by correct choice of the wind ratio, which remains constant in this traverse mechanism throughout the process. The wind ratio prevents the number of coils per double traverse, reaching a whole number effectively from empty to full package. Consequently, no ribboning or patterning is produced on the package. The wind angle becomes progressively narrower as the package diameter increases.

A winding machine based on either of the two principles, namely surface drive or spindle drive, can be equipped with any of the three traversing systems.

Consider now two successive layers of yarn on a package (Fig. 1.23). It is observed that the coils of one layer cross the coils of the other layer. The included angle between two crossing coils can be termed as the angle of wind or the wind angle. The normal range of angle of wind is from 30 to 55°.

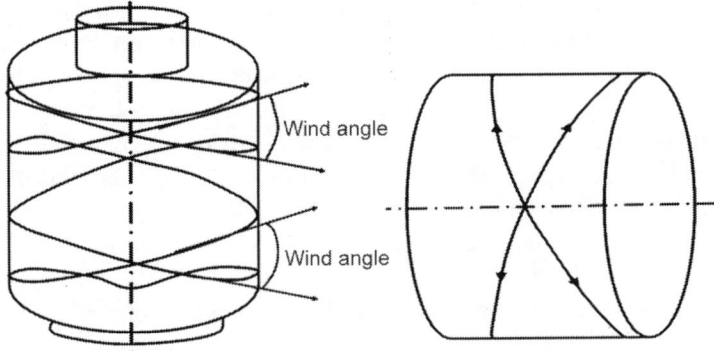

Fig. 1.23 Two successive layers on cross wound package

Minimum package density is reached when the wind angle is 90° or the two crossing coils are perpendicular to each other. Under such a condition, very large air gap would exist between adjacent crossing yarns. This degree of openness is not required even for dyeing packages for which an angle of 55° is quite adequate. Packages for warping and shuttle-less weaving are usually wound with an angle of 30°. As the wind angle changes from 30° to 55°, there is normally a reduction in package density by 20–25%. Cotton packages for dyeing usually exhibit a density of 0.30–0.35 g/cc, whereas those for warping and weaving have values between 0.40 and 0.50 g/cc. Besides affecting density of the package, the angle of wind also affects its stability. A cross-wound flange-less package owes its stability to the axial components of yarn tension, directed opposite to each other in two neighboring layers (Fig. 1.20).

In the event of any impactful axial disturbance, these oppositely directed forces counter the tendency of any layer to slip out of the package (slough-off).

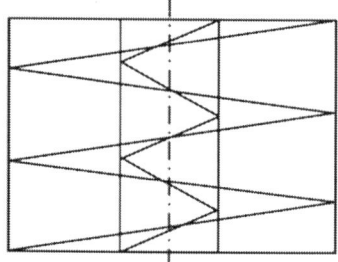

Fig. 1.24 Spindle driven package

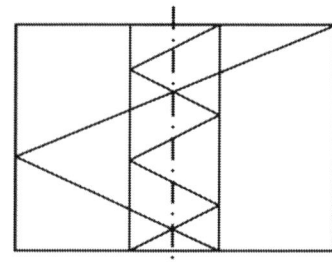

Fig. 1.25 Surface driven package

Evidently, a higher angle of wind would result in a larger axial component and hence a greater stability. A very low hypothetical angle of wind of say 5–10° would result in near parallel coils, resulting in very low stability. In case of pirn winding machine, low wind angle is observed. During shuttle checking, a pirn of 30 g yarn content can be subjected to an axial effective force of over 5 kg. This can disintegrate an incorrectly built pirn. A typical yarn layer on a pirn may contain 13 coils within a space of 3.75 cm. If winding was carried out throughout the pirn in this fashion, then a wind angle of about 3° would result on the shoulder and 9° on the nose. The wind angle of a coil on a package can be calculated from the formula:

Wind angle = $2 \tan^{-1} (1 / \pi D w)$

Where D, w, and 1 represent the package diameter, number of coils per traverse, and the length of traverse, respectively. Reducing the number of coils by a factor of 2 would raise the angles to 5° and 14°, respectively. It is inferred from these figures that coils on the shoulder of a pirn are more prone to sloughing-off, than those near the nose. However, this sloughing-off tendency can be suppressed by periodically switching over to smaller number of coils per traverse, aiming at a compromise between maximizing the capacity of pirn and minimizing loss due to sloughing-off. This is realized in practice by resorting to winding and binding layers. This concept is implemented practically by restoring to winding and binding layers on pirn surface during forward and backward journey of traverse guide, respectively.

A pirn winder is a spindle-driven machine in which the rotational speed of the pirn and the traverse speed are constant. Cylindrical packages like cheese can also be wound on such machines, especially when yarns with delicate surface have to be prepared. In such systems, the wind angle goes down with increase in diameter of the package but the number of coils wound per traverse remains constant (Fig. 1.24). Owing to the continuous drop in wind angle and instability of the package with increasing diameter, spindle-driven systems are not suitable for large diameter packages.

The surface speed, wind angle, and traverse speed of surface-driven winding machines remains constant. However, the number of coils per traverse continuously diminishes. Thus, very large packages can be constructed on surface-driven winders with appropriate stability. The continuously diminishing number of coils per layer with increasing package diameter becomes the cause of two critical problems, namely patterning and non-uniformity in package density.

The wind angle and the package density on a surface-driven winder with rotary traverse, i.e. employing grooved drum, is controlled by varying the scroll. Drums are usually made with 1.0, 1.5, 2.0, and 2.5 scrolls. With the 1.0 scroll, the drum makes one revolution as the yarn makes one complete traverse resulting in the lowest density and highest angle of wind. With the 2.5 scroll, the drum makes 2.5 revolutions for one traverse of the yarn. For a drum of diameter 'd', traverse 'l' and scroll 's', the theoretical crossing angle of the grooves is:

Crossing angle = $2 \tan^{-1} (l / \pi. d s)$

The corresponding expression of wind angle on the package would be

Wind angle = $2 \tan^{-1} (l / \pi. D w)$

Where, w represents the number of coils per traverse.

Principally, for surface driven winders, the product $D \times w$ is constant and is exactly equal to the product $d \times s$

$D \times w = d \times s$

Uniformity of package is another criterion, which is influenced by the nature of winding system. If a constant yarn length per unit surface area is decided as the measure of uniformity, then it can be shown that a continuously diminishing wind angle with increasing package diameter such that

D sin θ = constant,

where θ is the semi-wind angle, would satisfy the requirement. This condition is totally violated in random winders, where the wind angle is constant. Such packages would get progressively softer as one move from core to the surface. In precision winders, the wind angle does go down with increasing diameter; but the relationship is

D tan θ = constant.

This too does not quite fulfill the requirement. It does appear therefore that building a large and uniform package still necessitates conceptually new solutions.

The number of package revolutions in one traverse equals the number of coils in the corresponding layer. Hence, the quotient k between the package rpm and the number of double traverses per minute would yield the value of number of coils in two layers that is in one double traverse (traverse ratio). On a spindle driven winder, the package rpm remains constant; hence the number of coils in one layer also remains constant. But on a surface driven winder, the package rpm diminishes continuously as the package builds up; hence the number of coils in a layer also goes down with increase in package diameter. Therefore the number of crossings between yarns of two adjacent layers would remain constant in a package built on a spindle driven winder, whereas the same would keep on falling in a package built on a surface driven winder (Fig. 1.25).

Fig. 1.26 Ribbon or patterning on cone surface

If the total number of coils in two consecutive layers, i.e. the ratio k, is a whole number, then the yarn returns to the same winding point at the start of the next double traverse leading to patterning (Fig. 1.26). By careful selection of this number, it is then possible to shift the winding point at the beginning of each double traverse and introduce a phase shift amongst pairs of yarn layers around the package axis, distributing the crossing points in the process evenly over the package surface, provided this number can be maintained at a constant value. This is possible in spindle driven winders, which are therefore also termed as precision winders. The term "Gain of Wind" is employed to indicate the amount by which the wind per double traverse should differ from the whole number, so that not only a pattern is avoided but each yarn coil is also laid at a precise distance from its neighboring coil. In surface driven winders, such a control is not possible as result of which the yarn coils are located randomly on the package surface (Fig. 1.23).

Anti-patterning motions can be useful to eliminate the problem of ribbon formation on surface driven winders. Many commercial solutions are available in the market, but the concept of step precision winders and computer-aided package (CAP) is quite reliable. Anti-patterning devices are worked to avoid the critical values of K during package winding. Generally, the value of K keeps on falling due to a continuous drop in package rpm by increasing its diameter. The first principle solution is to reduce the traverse speed continuously and proportionately, so that a particular value of K can be maintained over a period of time during which the winding system is passing through the critical values of K as shown in Fig. 1.27. In coming layers, the traverse speed may be raised back to the original value in one step. This is quick climbing down of one value of K to another value. Such manipulation in traverse speed can be possible with the aid of microprocessor-controlled inverters.

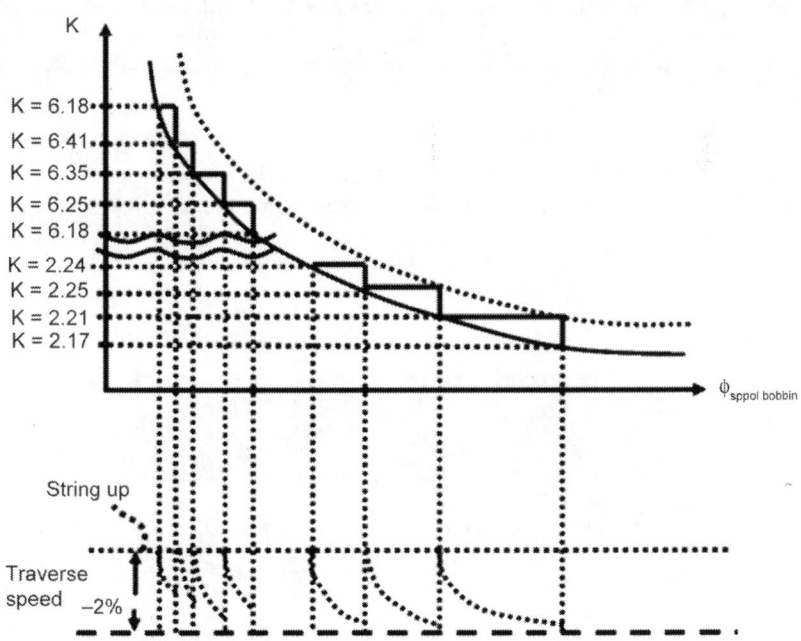

Fig. 1.27 Coiling on surface driven package

Principle of function of propack system

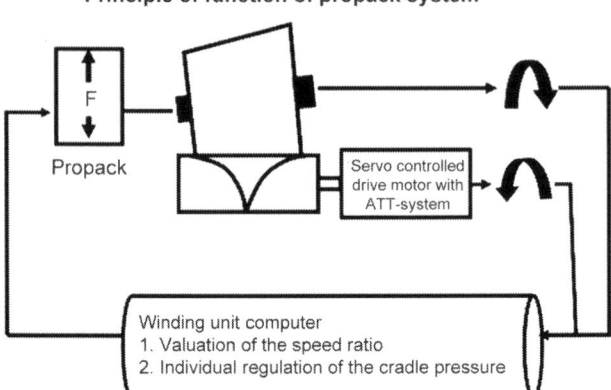

Line diagram based on saurer/schlafhorst concept

Fig. 1.28 Principle of propack system

The computer-aided package building principle is based on the observation that the effective contact point between a cylindrical drum and a conical package, which is located approximately at 1/3 the traverse distance from the package base, determines instantaneous package rpm. As and when the ratio of the package rpm and the traverse speed reaches a critical value favorable for patterning, the cradle holding the package against the drum is slowly rotated resulting in a gradual shift of the contact point from the base to the nose (Fig. 1.28). This process continues during the critical phase of pattern formation, holding thereby the instantaneous value of K at a non-critical level. Once this phase passes, the cradle is swung back to the normal position, re-establishing thereby the original contact point between the cone and the package (Fig. 1.29).

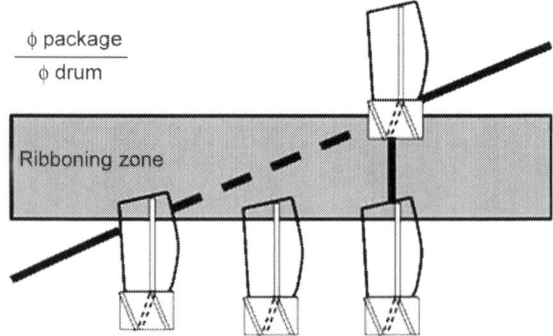

Fig. 1.29 Cradle swing mechanism to avoid ribbon formation

1.9   Tensioners

Certain level of tension is essential to be imparted to the running yarn to wrap a compact package and to enable the weak places to break. Tension uniformity is essential to maintain homogeneity in package density (Ukey et al., 2010). To achieve a cost effective quality in weaving preparatory, the role of tensioners becomes very critical (Latzke 1979). To achieve uniform tension, various tensioners are used on a winding machine. In the context of textile industry, tensioner is a device which is used to impart or enhance the tension on running yarn up to a prescribed limit. Tension-measuring devices, compensator and tension-generating devices are classified as shown in Fig. 1.30.

The tensioner should have the following features:
- They should be easily set to give the required tension within narrow limits.
- They should be self-cleaning.
- Tensioner should be free from dust and fly in case of spun yarn processing and oil in continuous filament yarns.
- They should not disturb the twisted structure of yarn.
- They should be centrally controlled.
- They should be self-threading.
- They should be cost effective.

The winding tension affects the following tensile properties:
- Tenacity
- Elongation
- Work- to-break

But excessive tension in winding will deteriorate the above-said tensile properties.

It is proved that the yarn tension is very important factor in textile industry and needs to be checked regularly in winding, warping, and sizing processes, where a high speed control of yarn tension is required. Variation in yarn tension arises various package faults.

There are two ways of measuring yarn tension in textile industry.
- Measuring with hand-held tension meters
- Using an active device that measures a feedback of the yarn tension in real time using sensors

1.9.1   Classification of tensioning devices

The purpose of tensioning device is to allow the maintenance of proper tension in the yarn in order to achieve a uniform package density.

The tensioners fall into three categories (Brunnschweiler et al. 1959)
- Capstan or multiplicative tensioner
- Additive tensioner
- Combined or disc tensioner

The principal characteristic of a tensioning device is its degree of control of output tension for varying input tensions for different unwinding tensions from the supply package. All yarn-tensioning devices consist either of two types of tensioning elements or combinations of both types.

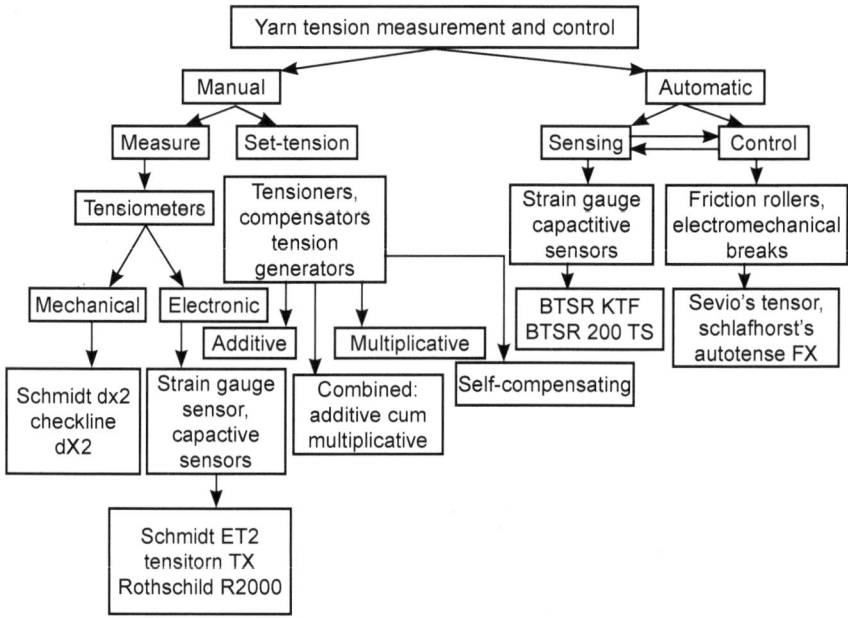

Fig. 1.30 Tensioners used on yarn winding machine**

Post or capstan or multiplicative tensioner
In this type of tensioner, a yarn is wrapped round or deflected by a post, giving an output tension which is multiple of the input tension. The output tension in this type of tensioner depends on coefficient of friction (μ) between yarn and post and angle of wrap of yarn on capstan denoted by θ (Jung et al., 2004). If the incoming tension is T_1 and outgoing tension is T_2 in Fig. 1.31, then

$$T_2 = T_1 \left(e^{\mu_1 \theta_1} + e^{\mu_2 \theta_2} + e^{\mu_3 \theta_3} \right)$$

$$\mu = \mu_1 + \mu_2 + \mu_3$$

$$\theta = \theta_1 + \theta_2 + \theta_3$$

If all posts are made of same material, then
$$\mu = \mu_1 = \mu_2 = \mu_3$$
If yarn is making same angle of wrap in radians around all posts, then
$$\theta = \theta_1 = \theta_2 = \theta_3$$
The outgoing tension is T_2
$$T_2 = T_1 e^{\mu\theta}$$
Where e = 2.718

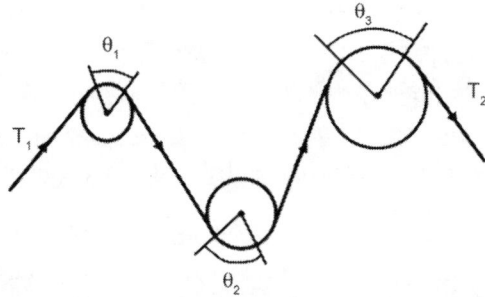

Fig. 1.31 Post or capstan type tensioner (multiplicative type)

The following facts are related to capstan-type tensioner:
- Since μ, θ and e are constants, the outgoing tension is merely a constant multiple of the incoming tension. Hence, its name multiplicative tensioner is justified well.
- If the incoming tensioner is zero, outgoing tension will also be zero.
    - To vary the tension, at least one of the following must be done:
    - Change the angle of wrap
    - Change the number of posts
    - Change the incoming tension

Additive type tensioner

In additive-type tensioner, the yarn passes through two plates in which one plate is base plate and the other is weighing plate as shown in Fig. 1.32. The outgoing tension in additive-type tensioner depends upon the coefficient friction between the weighing plates and the yarn (μ), and the force applied to the yarn by weighing plate is F
$$F = mg$$
The relationship between incoming and outgoing tension in an additive-type tensioner is
$$T_2 = T_1 + 2\mu F$$
$$T_2 = T_1 + 2\mu mg$$

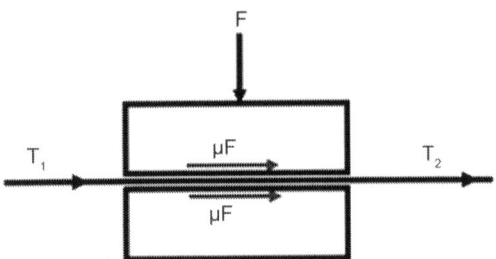

Fig. 1.32 Additive type tensioner The following observations may be made regarding additive type tensioner:

- Since μ, F and 2 are all constants for a given system, the outgoing tension is simply a constant added to the incoming tension, hence the name is additive.
- If the incoming tension is zero, there is still an outgoing tension.
- The outgoing tension may be changed simply by changing the loading F through mass of the upper plate M.

Combined tensioner

This is the most common tensioning device found on warping machine. This type of tensioner consists of a post and a disc. The yarn passes round a capstan post and at the same time passes between springs or dead weight loaded discs mounted on the post or capstan as shown in Fig. 1.33.

The discs mounted over post provide the effect of additive type tensioner, while the partial wrap of yarn around post adds multiplicative effect on outgoing tension. Both additive and multiplicative effects are shown in Fig. 1.34.

$$T_2 = T_1(e^{\mu\theta}) + 2\mu F$$

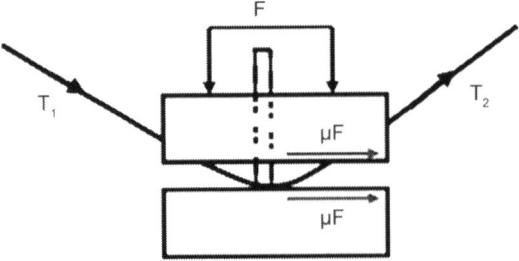

Fig. 1.33 Combined type tensioner (side view)

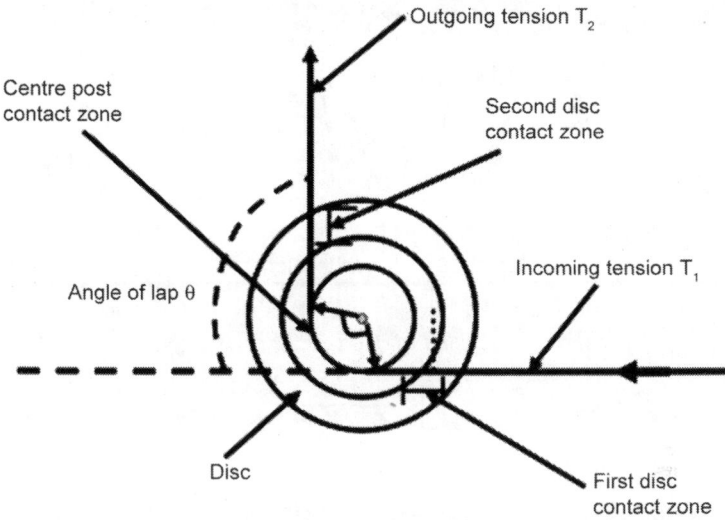

Fig. 1.34 Combined type tensioner (top view)

The following observations may be made regarding combined type tensioner:

- Since μ and 2 are constants for this system, the outgoing tension can be tuned by both playing with F and θ by alteration of number of dead weights on discs and angle of lap around post, respectively.
- If the incoming tension is zero, there is still an outgoing tension.

1.10   Different mechanisms to impart tension

The conventional way of imparting tension on running yarn is dragging the yarn under a dead weight. Generally, these dead weights are ball or disc kept in an appropriate frame as shown in Fig. 1.35. The easiest way of imparting the yarn tension is to deflect the yarn path around single or multiple stationary posts. The outgoing tension is decided by the angle of wrap the yarn makes around the posts and the number of posts in the passage. This principle of deflection is utilized in grid bar method or gate type tensioner. In this method, the yarn passes between two series of grid posts in which one series being stationary and the other swinging between these fixed grid bars as shown in Fig. 1.36. The resultant tension is adjusted by changing the position of moveable grid bars relative to the fixed grid bars, either by spring pressure or weight and lever arrangement or by magnetic arrangement.

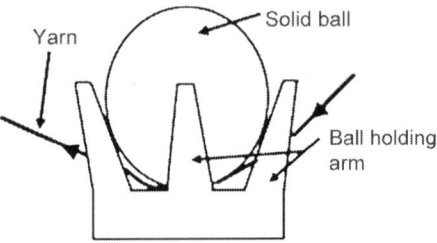

Fig. 1.35 Ball type tensioner

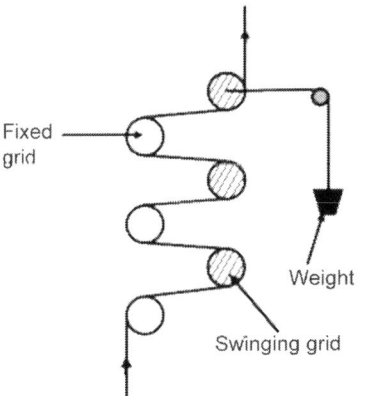

Grid bar or gate type tensioning device

Fig. 1.36 Gate type tensioner

1.10.1   Performance of various tensioning devices

The performance of various tensioning devices can be described by the classical theory of friction, and pictorially illustrated in Fig. 1.37 (Eiichi 1964).

(a) Indicates the outgoing tension when tensioning device has zero effect on yarn, and acts as a line of reference.

(b) Shows that outgoing tension is purely multiplicative or post tensioning. The steepness of this line varies with the contact angle of yarn around capstan or post and the coefficient of friction between yarn and post material.

(c) Shows that outgoing tension is purely additive or disc-type.

(d) Shows that outgoing tension is based on combined type tensioner, obtained from commercial combined disc and post or capstan. This type of tensioner causes divergence from line (B) and (C).

(e) Shows the outgoing tension of a compensating or gate type tensioner. The outgoing tension remains constant only up to a point when incoming tension does not exceed a certain value. The inclined portion of outgoing tension indicates about the ineffectiveness of tension-compensating system, and post-tensioning system is being produced from the various guides of tension compensating systems.

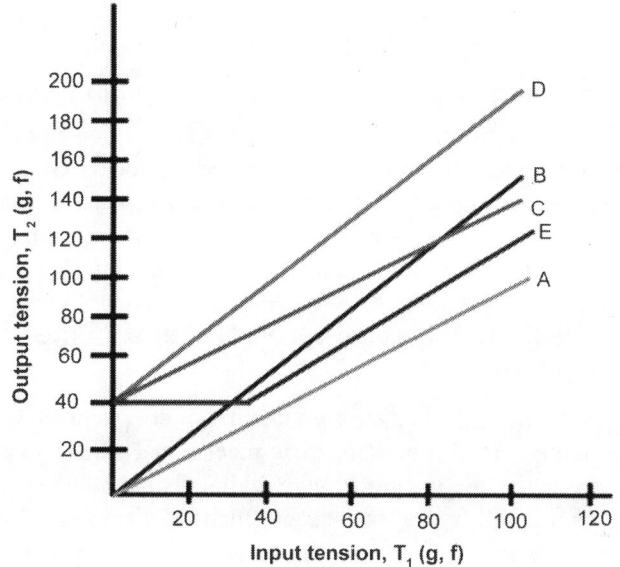

Fig. 1.37 Performance of various types of tensioners

1.10.2   Tension measuring techniques

In textile industry, the yarn tension measurement is performed by tension meters in dynamic conditions while online tension measurement in laboratories for research purposes. Various sensors are used for tension measurement of running yarn in case of electronic-type tensioner like Rothschild electronic tension meter (Bhattacharyya et al., 2003).

On the basis of operating principle, yarn tension measuring system can be categorized as

- Mechanical type
- Electronic type
- Combination of electronic and mechanical type (electromechanical type)

1.10.3   Mechanical type yarn tension measuring system

The mechanical type of tension measuring devices use mechanical parts such as pulleys and the rollers, and are generally used in measuring yarn tensions in processes such as weaving and warp preparation. These tension-measuring systems are free from the use of active sensors or electronics. Passive elements like feelers and springs are used to feel the tension in running yarn. Schmidt's mechanical hand held tension meter and Checkline's tension meter are mechanical tension meters.

1.10.4   Electronic-type yarn tension measuring system

Electronic type yarn tension meters are able to display tension amount in digital form which gives ease of tension adjustment during process. Strain, Gauge or Hall effect sensors are used to develop these types of tension meters. These tension meters are very precise in measurement, hence used in winding, twisting, and now in sizing processes where online tension measurement is beneficial.

1.10.5   Electromechanical-type yarn tension measuring system

These groups of tension meters are combination of electronic and mechanical type tension meters. In this system, basic mechanical parts like pulleys and rollers are used which are integrated with digital display of tension.

By comparing all types of tension meters, electronic yarn tension measurement gives an edge over other measurement systems due to following reasons:

1. Very precise tension measurement is possible at high velocities.
2. Wide range of tension measurement is possible.
3. Connectivity to computers is very easy.
4. Data can be stored and statistics can be applied.
5. Best frequency response.
6. Mobility of the sensing heads becomes very simple.

1.11   Package faults

A faultless package would contain a definite length of yarn, free from objectionable faults, wound on a suitable tube in such a manner that the yarn can be fully utilized in the subsequent process without causing any interruption during further manufacturing process. A pirn, for example, should contain a very large length of yarn that should guarantee a minimum number of pirn changes during weaving. This will not only lower down the fabric defects associated with pirn changes, but also contributes to better efficiency by reducing loom stoppages due to pirn change. The pirn should unwind smoothly during the picking process without disintegrating. Moreover, the yarn in the pirn should be sufficiently strong (free from weak places) and free from any faults such as slubs, big knots, undesirable hairiness, etc.

Few types of faults are peculiar to specific package form. Thus, the dimension of the pirn such as its diameter should match the shuttle exactly; otherwise a perfectly prepared pirn might be a complete waste just because it does not fit the shuttle. The tolerance here is clearly very low. But the dimensions of a cheese or a cone are not so critical for the subsequent processes. Hence pirn diameter and uniformity of the same throughout the package length needs to be closely monitored, strict specifications set up and observed. Such a regime would not be necessary for a cone or a cheese for which strict specifications and control would be needed for other factors such as package density, package hardness, package weight, length of yarn in the package, etc.

Notwithstanding the dissimilarities in emphasis on types of faults between packages of different types, some commonly encountered faults in pirns, cheeses, and cones are listed in the following:

1.11.1   Package shape

As mentioned in the foregoing, the exact shape of a pirn (Fig. 1.38), described by its diameter, chase length, chase angle as well as the full length of the pirn, determines not only its contour and its compatibility with the shuttle but also the amount of yarn that it would contain. The chase lengths can differ near the nose, near the back, and in the body (Fig. 1.23). Low chase angles (6–10°) are prescribed for continuous filament yarns, whereas angles up to 15° may be used for cotton yarns. Pirns containing filament yarns are usually wound to a smaller diameter and larger chase length as compared to those containing cotton yarns. For controlling these variables properly, the traverse mechanism needs to be set correctly as also the racking mechanism. Worn out gears become the cause of ribbon formation on pirn due to unequal shift in traverse. Fluctuation in winding tension can also lead to improper package shapes such as bulging of package.

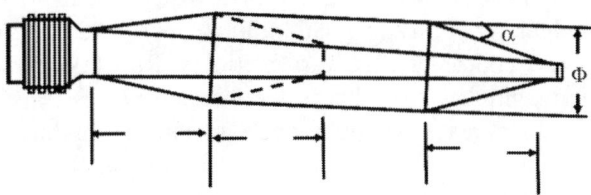

Fig. 1.38 Package shape

1.11.2   Length variation among wound packages

High length variation among packages results in frequent machine stops (as in warping), generation of excessive knots, and unwanted rewinding of unequally sized leftovers. If these leftovers are rewound and used in warp or weft or both, then depending on fabric quality, a varying degree of drop in loom efficiency would be observed.

1.11.3   Knots

A weaver's beam having a large number of knots gives poor performance during weaving. Knots detract from the general fabric appearance also. Quite a number of these knots can be traced to the supplied wound packages. End breakage during warping or weaving would also result in knots in fabrics. In modern winders, different types of splicers are provided for joining the broken yarn ends. A spliced joint is usually up to 1.2 times the thickness of the parent yarn and about 80–90% as strong.

1.11.4   Hairiness

Hairiness refers to the degree to which a yarn has fibers or "hairs" protruding from its main body. Yarn spun from staple fibers can never be free from a certain degree of hairiness. The characteristic of the yarn that refers to the number and extent of protruding fiber ends or loops is termed as hairiness or hairiness index. Some amount of hairiness is generally acceptable, as it helps in trapping air near the fabric surface and helps in increasing the thermal insulation of fabrics. It also provides a soft feel. However, for normal fabrics and yarns, excessive hairiness is known to detract from the fabric appearance, especially of fabrics with open structures. Moreover, excessive hairiness leads to poor shed formation during weaving. Long hairs can also roll into neps due to the yarn rubbing against machine parts. During winding, the hairiness of yarn goes up on account of excessive rubbing with worn out contact surfaces as also due to higher tension. Hairy yarns have detrimental effects on printing quality.

According to Barella (1957), yarn hairiness contributes all three protruding fiber ends, fiber loops, and wild fibers. Wang revealed that 82–87% protruding fiber ends (including both leading and trailing ends), 9–12% fiber loops, and 4–6% wild fibers contributed the yarn hairiness. Rust (1992) explored that fiber migration continues even during the winding process. Increasing fiber migration leads to increase in yarn hairiness and higher winding tension, and/or higher yarn velocity also leads to more fiber migration and hence more severe yarn hairiness. During the yarn winding, as a yarn passes the tension disk, the yarn guide, and the grooved drum, the loose fiber ends on yarn surface are rubbed severely, which inevitably causes alteration in yarn hairiness. At optimum twist factor, the pulling of fibers becomes very difficult to enhance yarn hairiness.

1.11.5   Package density variation

Package density variation is a major problem in case of drum winding machine. Due to package density variation, shape of the package is also distorted as shown in Fig. 1.39.

With coarse counts, where big packages are very important, variation among package densities is a major problem.

Density directly affects the outer diameter of a package, and in any warping creel there is a limitation on maximum package diameter. As an example, on a typical warping creel with a maximum possible package diameter of 255 mm, a package weighing up to 3.35 kg is workable. However in a 3 kg package, the package diameter may vary from 240 mm to 260 mm. But if this variation is reduced, then the same creel may be able to handle even a package of 3.5 kg, leading to higher production. In addition to this factor, density variation also causes difficulty in calculation of set length. It would also be difficult to ship packages of different densities in standard cartons, as the space requirement would vary. Density variation within package can cause fluctuation in unwinding tension and also may lead to instability, in addition to resulting in uneven shade when dyed in package form.

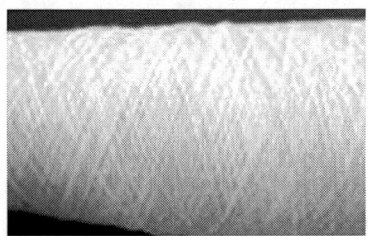

Fig. 1.39 Package with density variation

1.11.6   Wild yarn or foreign yarn

Normally it generates during splicing or knotting on a winding machine. Due to this problem, multiple breaks happen during warping or sizing and also on occasions during weaving. Further, use of such packages as weft would result in weft stops on loom during weaving. Frequency of wild yarns is higher in case of finer yarns.

1.11.7   Slough off

Slough off occurs when, during unwinding of a package, a bunch of coils slips out. Improper build as well as low winding tension may cause such problems in a pirn. As far as a cheese or a cone is concerned, patterning might be an additional source for such faults.

1.11.8   Cobweb

Cobweb is a yarn chord on the right hand or left hand flank of the cone or a length of yarn exiting from one of the package flanks and getting wound around the protruding end of the cone/sleeve and re-entering the flank at a different place. The source of such faults can be improper cradle, functioning of the drum or improper yarn tension.

1.11.9   Defective /damaged tube

Due to this problem, higher breaks occur during unwinding of the innermost yarn layers from the package. This happens mainly due to yarn entanglement with the damaged surface of the tube. For pirns, different types of surface are prescribed for yarns of different nature so that the coils are gripped properly.

1.11.10   Damaged package

Damaged packages such as cut cone or cheeses or crushed surface of pirn are major faults that can be minimized by taking better care in packing, package selection and transportation.

1.11.11   Cauliflower

On the nose flank of the package next to the cone/sleeve, yarn layers are pushed out and tend to curling. This is due to high pressure of the outer yarn layers upon the package core.

Yarn waste in the cones
This is due to loose yarn ends that are wound on to the cone.

Stitch, drop over, web
Yarn is visible on the small or on the big side of the cone either across the side, around the tube, or going back in the cone.

Damaged edges or broken ends on the cone
The yarn is broken on the edges or in the middle of the cone.

Ring formation
The yarn runs in belt formation on to the package, because it is misguided.

Without transfer tail
The desired transfer tail is missing or too short.

Ribbon formation
Pattern or ring formation is made by the drum when rpm is staying the same.

Displaced yarn layers
Yarn layers are disturbed and are sliding towards the small diameter of the cone.

Misguided yarn
The yarn is not equally guided over the whole package.

Cauliflower
On the smaller side of the package, the yarn shows a wrinkle effect.

Soft and hard yarn layer
Some layers of yarn are pushed out on the small side of the cone.

Soft and hard cones
Great difference in package density from one winder head to another

1.12   Knotting and splicing

Joining of broken yarn ends on a winding machine is a basic requirement to prepare yarn packages of desired length, shape, and size for efficient working of the subsequent processes like warping, sizing, and weaving. Earlier, knotting was the only process for joining of yarns together; but now this is being gradually replaced by splicing. Knot itself is an objectionable fault due to its physical dimension, appearance, and problems during downstream processes. A high quality yarn cannot be produced without replacement of knots. Splicing is the best option to eliminate the problem of knotting and piecing; although, the tensile strength of yarn with knot is superior to that of the yarn with splice.

1.12.1   Knotting

Knots are used to provide continuity to the two yarn ends at different stages in the textile industry. A wide variety of knots like dog knot, weaver's knot, double weaver's knot, Fisherman's knot, and quick knot, etc., are used in textile industry. Few basic knots are pictorially illustrated in Fig. 1.40.

1.12.2   Dog knot

The fastening of dog knot is very simple and quick, but its size is bigger than Fisherman's and weaver's knot. Dog knot is very much suitable for creeling, but it can be incorporated into fabrics.

1.12.3   Fisherman's knot

The tying of Fisherman's knot is not convenient to tie by hand at all. This knot is very popular in textile industry particularly in cotton, woolen, worsted, blended, and filament yarn industry.

1.12.4   Weaver's knot

Weaver's knot is most popular in cotton and worsted textile industry. The diameter of weaver's knot is 2–3 times that of the parent yarn. Both single and double weaver's knots are frequently used in textile industry, depending on the requirement of knot slippage and slippery properties of the parent yarn.

The resistance to slippage (knot slippage or passing through reed and heald eyes, etc.) for three basic knots is as under

Fisherman's knot < Weaver's knot < Dog knot

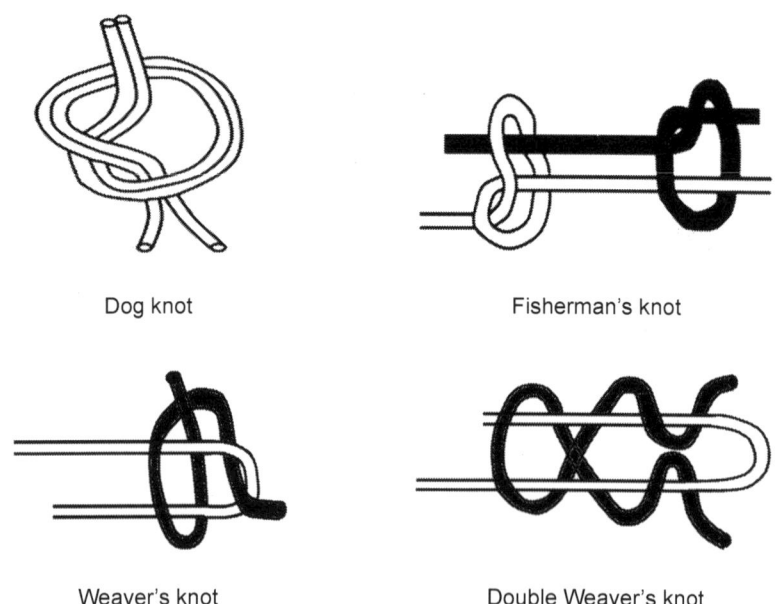

Dog knot   Fisherman's knot

Weaver's knot   Double Weaver's knot

Fig. 1.40 Few basic knots

Above three knots can be tied by hand or hand knotter or by automatic mechanical knotter. Use of mechanical knotters also offers advantages like uniformity in knot size and tail ends. Automatic mechanical knotters are available on winding machines which are dedicated to process only specific types of yarns, which are difficult to splice or whose splice performance is very poor. Automatic knotters are available on winding machine in two forms.
- A knotter for a group of winding heads
- A knotter dedicated to each winding head

1.12.5   Quality of knot

Assessment of quality of knot is essential because it contributes to process failure in winding and end breakage in subsequent processes. The quality of knot is evaluated in terms of
- Slippage tendency
- Knot size
- Length of tail ends
- Slippery tendency of knot against different surfaces

The best quality of knot is that which gives least breakage and slippage under weaving conditions.

The knots are found to become the cause of warp breakage in weaving in three ways.
- Size of knot being too big to pass through drop wire, heald eyes, or reed dents
- Due to inadequate slippage of knot
- Long tail ends of knot getting entangled with adjacent warp threads
- Factors which affect the slippage of knot are
- Coefficient friction of the material of the yarn
- Twist, count, and surface coating of the yarn, if any
- Geometry of the knot, i.e. number of crossing points, angle of wrap at crossing point

Basic properties of yarn and geometry of knot decide the possibility of slippage of knot. Length of tail ends is also an important factor. Shorter tail ends of knot are liable to slip easily during weaving. The tail's length should basically be judged from the linear density and flexural rigidity. In yarn with lower linear density or lower bending rigidity or both, failure of knot (knot slippage) starts in case of shorter tail ends and vice versa for yarns with greater linear density and bending rigidity (flexural rigidity). The length of tail ends should be 5–10 mm in weaver's knot, while 6–12 mm in Fisherman's knot depending on slippage resistance of yarn and knot. Direction of tail ends also plays an important role in knot failure.

- In weaver's knot both tail ends point in same direction, while in Fisherman's knot the direction of the tails are opposite to each other.
- The thickness of Fisherman's knot is the sum of the diameter of the yarn and the knot, whereas the thickness of weaver's knot is that of the knot itself.
- The tension of loom is of cyclic nature and tends to loosen the knot.
- The slippage resistance of knots in spun and single yarn is higher than smooth multifilament and double yarn.
- If the knot is not strong and flexible enough, it is more prone to either slip or break under the action of large transverse forces developed in the knot when knotted yarns are stressed.
- The strength and flexibility of a knot are represented by an index called "knot efficiency".

$$\text{Knot efficiency} = \frac{\text{Strength of knotted portion of yarn}}{\text{Strength of parent yarn without knot}} \times 100$$

- In case of conventional winding machine, the quality of knot depends entirely on the winder's skill and operating skills.

1.12.6   Disadvantages of knots

- With high density woven fabrics, knots and its tails act like a scissor on neighboring ends and also hamper the shed formation. They also strike against thread and cause yarn breakage.
- The diameter of Fisherman's knot is 3–4 times and that of the weaver's knot 2–3 times more than the yarn itself. The bulkiness of the knot and knot tails are friction points during withdraw of the yarn over the ends of the package and may cause slough-off.
- The mass of knot disturbs the weft insertion on air-jet loom which results in loom stoppages and fabric fault at higher cost.
- When knots pass through a disc tensioner, a sudden increase in tension occurs which leads to process disturbances.
- The closed reed setting with plied yarns for terry cloth may cause a knot in front of reed to spring back and form loop.

1.12.7   Mesdan's knotters

In spite of the ever growing demand for knot-free joining technology, MESDAN® continues to manufacture a complete range of knotting devices in both manual and automatic versions. Three principal knot configurations are winders, weaver's, and fisherman's. These knotters still represent the only means to join in safe and repeatable way monofilaments, bonded yarns or any yarn which cannot be opened into fibers due to finishing.

1.13   Splicing

The next important element in the winding process is the splicer. Splicing is the technique of joining two yarn ends by inter-mingling the constituent fibers as shown in Fig. 1.41, so that joint is not significantly different in the mechanical properties and appearance with respect to parent yarn. Yarn splicer comes in varied forms for different types of yarns. A splicer may function on pneumatic, hydraulic, thermal or mechanical principles. Splicers are employed for joining single yarn spun from staple fibers or from continuous filaments. Splicers can also be employed for joining broken plied or even cabled yarns.

Concept of ideal splicing
An idealized splicing can be defined as under:
- Spliced portion should be as short as possible and uniform.
- Spliced portion thickness should not be significantly excessive over length than parent yarn.
- Spliced portion should be durable enough.
- Tensile strength of spliced portion should be comparable with parent yarn.
- Fibers at spliced portion must lie relatively parallel with their length distribution as regular as possible.
- The number of fibers in the two yarn ends should be same or about 10% lower than the normal cross-section.
- To attain the requisite strength, the spliced portion must have as much twisted as possible (twist may be more than mean twist of normal yarn).

To define idealized spliced structure, the spliced structure has been divided into three zones to make it convenient. Transverse and longitudinal studies revealed that the splice structure of yarn comprises of three different regions brought by wrapping, twisting, and tucking intermingling.

Wrapping
The tail end of each yarn strand is tapered and ends with few fibers. The tail end makes an ample wrapping of several turns and thus prevents separation of splice portion. Fibers of the twisting yarn are embraced to the body of the parent yarn and thus act as a belt. This in turn gives appearance to the splice portion of the yarn.

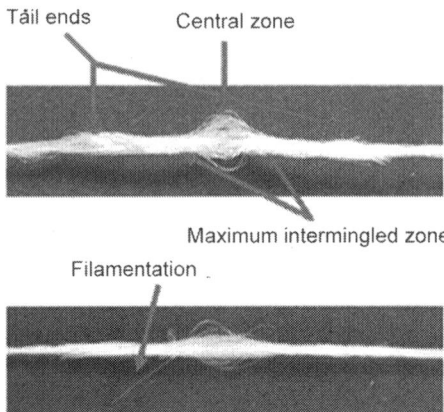

Fig. 1.41 Spliced yarn portion

Yarn twisting

The two yarn ends comprising the splice are twisted around the body of the yarn; each yarn strand twists on the body of the parent yarn on either side of the middle of the splice. The cross-section of this twisting region distinctly shows the fibers of the two yarn strands separately without any intermingling of the fibers.

Yarn tucking or yarn intermingling

The middle portion of the spliced joint is a region (2–5 mm) with no distinct order. The fibers from each yarn end intermingle in this splice zone, just by tucking. The studies on quantitative contribution of splice elements showed

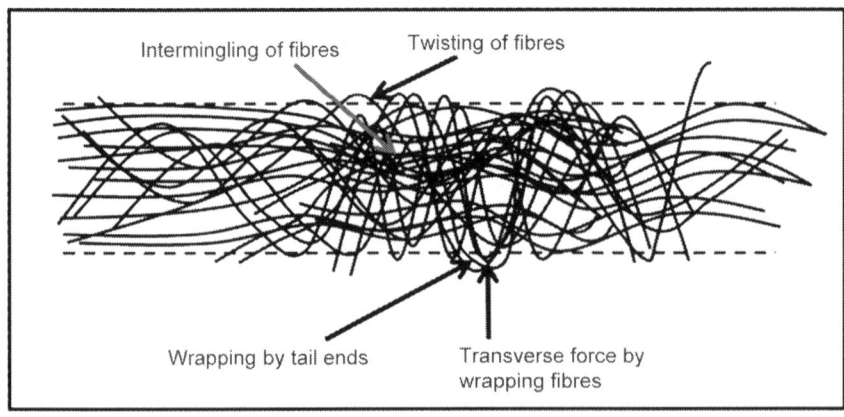

Fig. 1.42 Ideal fiber disposition in pneumatic spliced portion of spun yarn

that intermingling/tucking contributes the most to the strength of splice (52%), followed by twisting (33%) and wrapping (about 15%) as shown in Fig. 1.42.The lower strength of the splice is attributed mainly to the lower packing coefficient of the splice zone or lower tucking of fibers. Spliced yarn has a lower breaking elongation than normal yarn. Breaking elongation is mainly affected by intermingling. Wrapping and twisting provide mainly transverse forces. The absence of fiber migration gives lower breaking elongation to splice.

1.13.1   Quality assessment of yarn splice joint

Appearance and strength are two important characteristics of a splice joint. Although, quality of splice can be assessed by methods like load-elongation, work of rupture, % increase in diameter, and evaluation of its performance in downstream processes, etc. The appearance can be assessed either by simple visual assessment or by comparing with photograph of standard splice. Comparison of splice joints by image processing is quite quick and precise.

1.13.2   Types of splicing

Principally, yarn splicing can be classified into four distinct classes.
- Pneumatic splicing
- Mechanical splicing
- Electrostatic splicing
- Thermo splicer

Pneumatic splicing
Pneumatic yarn splicing is well-established technique for joining two yarn ends together by means of a blast of compressed air in a splicing chamber. The top view of splicing chamber is shown in Fig. 1.43. The thread ends to be joined are mingled and twisted together in spliced chamber. A typical splice chamber is shown in Fig. 1.43.

KZ automatic air splicer is the leading air splicer and supplied to Savio Shandong Textile Machinery Co Ltd. Savio incorporated KZ906Z on their very popular winding machine "Orion Super M/L". A wide variety of yarns can be processed on this splicer. The prime characteristic of KZ906L(690L) is that all setting parameters are set at the winder computer, thus speeding up the machine set up for a new yarn lot and eliminating undesired difference between the various heads. It is also possible to store the setting data of specific yarn in the winder computer and recall them when required. The KZ906L splicer is precision built and manufactured with the highest grade of material available, and is sealed against penetration of dust and fugitive fibers, the results of which is high performance splicing with minimal maintenance.

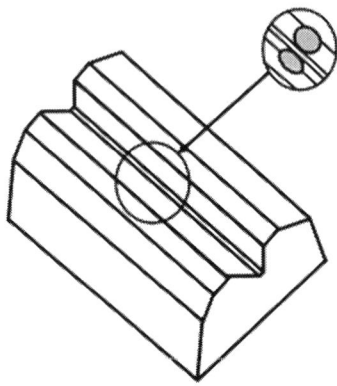

Fig. 1.43 Standard V-shaped splicing chamber

Principles of pneumatic splicing

The principles of splicing are simple to explain, although the process is quite complex. Direct observation of splice formation is difficult, because the process occurs extremely quickly. Even high-speed photography is only of limited use. Most of the conclusions about splicing performance have come from indirect observations. For example, by using tracer fibers, splicing performance can be observed and photographed. Understanding of the splicing process is therefore based on interpretation rather than theoretical analysis.

The process of splicing is easiest to understand in case of a simple transverse splicing chamber, with a single central blast hole, splicing a continuous filament yarn with a low twist level.

The yarns to be joined are placed into the splicing chamber. This is simplified by providing the chamber with a hinged cover (a chamber pad, or pad), which is normally open. When the pad is closed and the waste ends are cut to length by the integral knives, an air blast is introduced into the splicing chamber at very high speed as shown in Fig. 1.44. The air is highly turbulent, and the violent small-scale disturbances radically disrupt the arrangement of the fibers in the splicing chamber. Those fibers that happen to lie across the opening of the air-feed hole are separated by the direct blast. Those, which lie elsewhere in the chamber, are subjected to a chaotic pattern of vortices downstream of the entry point, which produce twisting and intermingling.

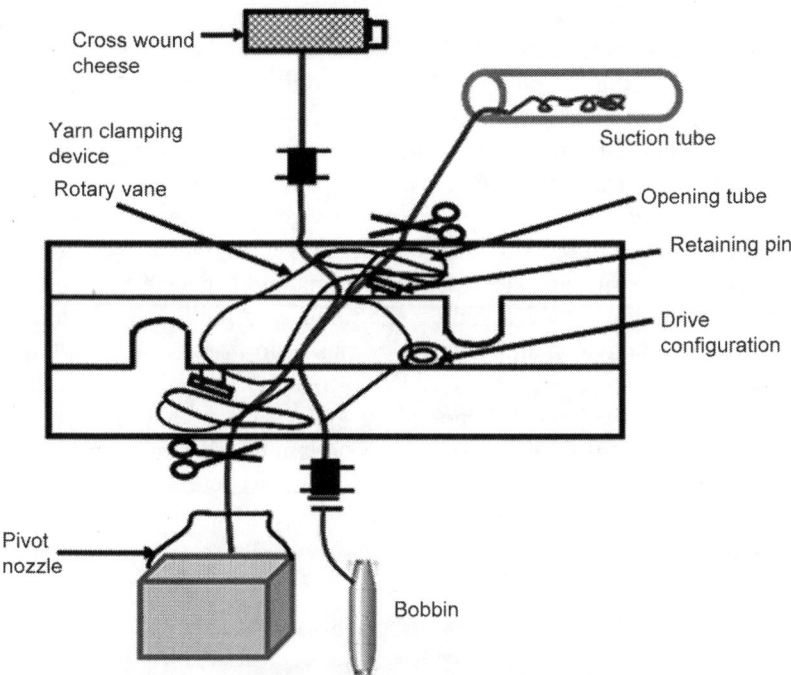

Fig. 1.44 Pneumatic splicing principle

The resulting splice has a characteristic form. The central section, which corresponds to the air entry point, appears essentially unchanged, with the fibers lying largely parallel. On either side of this central section, the fibers lie in dense clusters, highly twisted and intermingled together. Each cluster usually terminates in a small tail where the extreme tips of the spliced yarns have not been fully bound into the structure. When a load is applied to the splice assembly, the fibers in the clusters slip very slightly, until the entire structure stabilizes, as the inter-fiber frictional forces take the load.

Turbulence is random, so no two splices are identical on the micro-scale. Nevertheless, the whole splice is much longer than the scale of the intermingling, so splices for all practical purposes are identical. With continuous-filament yarns, using even a very simple splicing process, very high splice strengths can be achieved, typically 90–95% of that of the parent yarn.

Staple yarn comes in a multiplicity of forms: the staple length may be short or long; the spinning system may take one of many forms; the twist level can take almost any value; the yarn may be single, two-fold, or multi-fold, with almost any level of folding twist; the yarn may be assembled from one or many fiber types.

The splicing process acts at the level of the individual fibers. It is therefore clear that the splicing of spun yarns, though conforming in general terms to the simple account, must be more difficult. Much of the development effort required for making pneumatic splicing more universal in its application has gone into the generation of specific solutions for staple yarns.

For spun yarns, with their more complex construction, it is necessary to present the yarns to the splicing chamber in a form, which facilitates the splicing action. To this end, some splicer manufacturers have developed end-preparation techniques, which subject the fibers to a preliminary treatment. Others have used novel chamber forms as shown in Figs. 1.46–1.48. Others have modified the splicing process itself.

As a general rule, the appearance of splices made in staple yarns is rather less compact than those made in continuous filament yarns. If the yarn construction is particularly difficult, the splices may be fully acceptable in terms of mechanical properties, but may have fluffy "tails" as shown in Fig. 1.45.

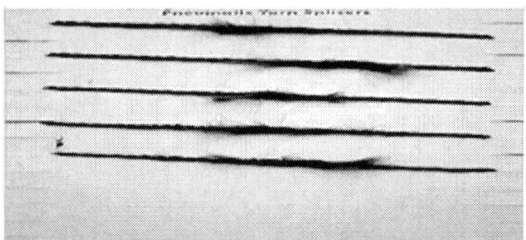

Fig. 1.45 Pneumatic yarn splice

The end preparation affects the splicing performance in many ways as shown in Fig. 1.46

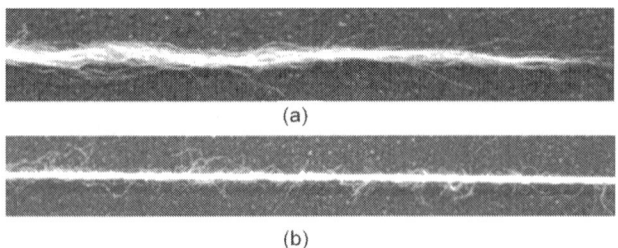

(a)

(b)

Fig. 1.46 (a) Optimum length opening end for splicing
(b) Optimum length opening spliced joint

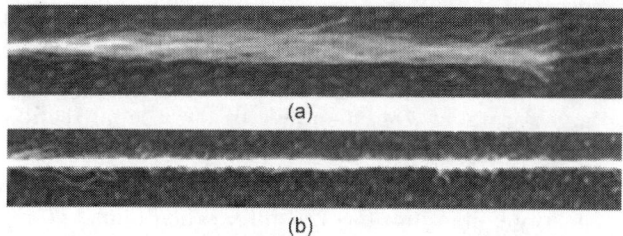

(a)

(b)

Fig. 1.47 (a) Short length opening end for splicing
(b) Short length opening spliced joint

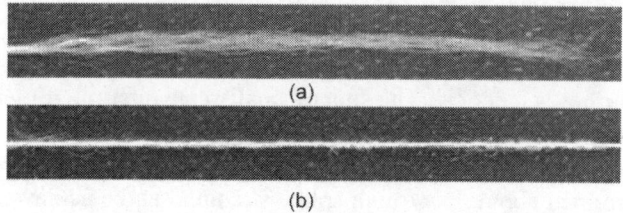

(a)

(b)

Fig. 1.48 (a) Optimum untwisted end for splicing
(b) Optimum untwisted spliced joint

Issa and Grutz (2205) demonstrated a method for optimizing the opening process on the splicer by which a loop former is added to form the loop outside the opening tube in order to prevent irregular opening and thinning of the yarn ends. The opening and splicing results obtained by using this principle were better than the results on the standard splicer.

The principle of pneumatic splicing can be understood by splicing a continuous filament yarn with a low twist level in a simple transverse splicing chamber, with a single central blast hole. The opposite end splicing is performed in a simplified hinged cover splicing chamber.

The first step of splicing is laying of both yarn ends in the splicing chamber, followed by pad closing and then cutting of extra tail ends by integral knife arrangement. After a very short duration of end trimming, an appropriate transverse air blast (transverse to the yarn ends) takes place into the splicing chamber. The intensity of transverse air blast remains very high. The behavior of such a compressed air blasting at very high speed remains turbulent. Any violation and small disturbances disrupt the radial arrangement of the fibers in the splicing chamber.

Those fibers which happen to lie across the opening of the air-feed hole are separated by the direct blast. Those which lie elsewhere in the chamber are subjected to a chaotic pattern of vortices downstream of the entry point, which produce twisting and intermingling.

When the air supply is cut off and the chamber is opened, the resulting splice has a characteristic and reproducible form. The central section, which corresponds to the air entry point, appears essentially unchanged, with the fibers lying largely parallel. On either side of this central section, the fibers lie in dense clusters, highly twisted and intermingled together. Each cluster usually terminates in a small tail, where the extreme tips of the spliced yarns have not been fully bound into the structure. When a load is applied to the splice assembly, the fibers in the clusters slip very slightly, until the entire structure stabilizes, as the inter-fiber frictional forces take the load.

A splice is produced by the reaction of fibers to turbulent air. Turbulence is by definition a random process, and therefore no two splices are structurally the same on the micro-scale. Nevertheless, the length of the splice, compared to the scale of the intermingling, is such that successive splices are for all practical purposes identical. The sum of small-scale intermingling results in a whole which is consistent from splice to splice.

With continuous-filament yarns, using even a very simple process such as that described above, very high splice strength can be achieved, typically 90–95% of that of the parent yarn. The advantages of splicing over traditional knotting are thus clearly demonstrable; the appearance of a splice is usually less obtrusive than that of the knot, and the strength is usually much greater.

The above is an idealized account of a splice of the simplest kind. The picture becomes much more complex when staple yarn is being spliced. Staple yarn comes in a multiplicity of forms: the staple length may be short or long; the spinning system may take one of many forms; the twist level can take almost any value; the yarn may be singles, two-fold, or multi-fold, with almost any level of folding twist; the yarn may be assembled from one or many fiber types.

The splicing process acts at the level of the individual fibers. It is therefore clear that the splicing of spun yarns, though conforming in general terms to the simple description outlined above, must pose a more taxing set of problems. Much of the development effort required to make pneumatic splicing more universal in its application has gone into the generation of specific solutions for staple yarns.

With spun yarns, with their more complex construction, it is necessary to present the yarns to the splicing chamber in a form which facilitates the splicing action. To this end, few splicer manufacturers have developed end-preparation techniques, which subject the fibers to a preliminary treatment. Others have used novel chamber forms. Few have modified the splicing technique itself, and it is this choice of splicing procedure which is described in another section.

Mechanical splicer

Mechanical splicers are used to splice heavy denier denim yarns, compact yarns and core spun elastomeric yarns. Twin splicer is a mechanical splicer as shown in Fig. 1.49. Rotating discs are used in mechanical splicer as shown in Fig. 1.50.

The splicing unit that could be incorporated into the winder and of particular interest was the Twinsplicer (Fig. 1.49), which is a mechanical system that can splice not only denim yarns but also compact and core-spun elastomeric yarns.

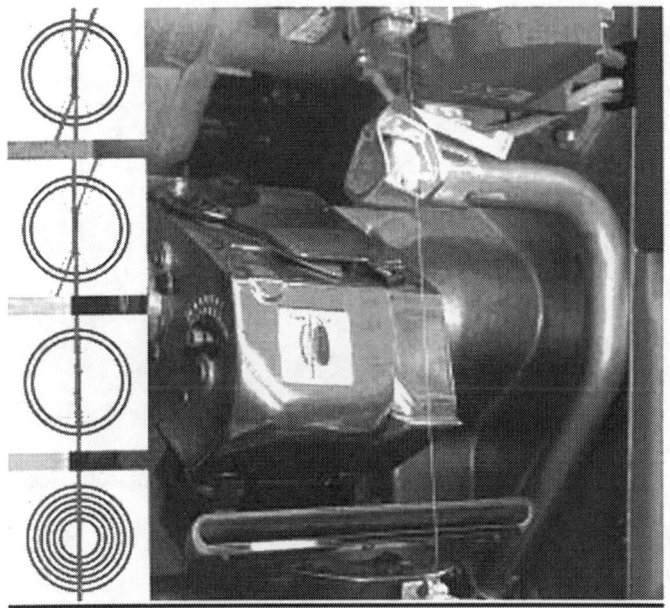

Fig. 1.49 Mechanical splicer

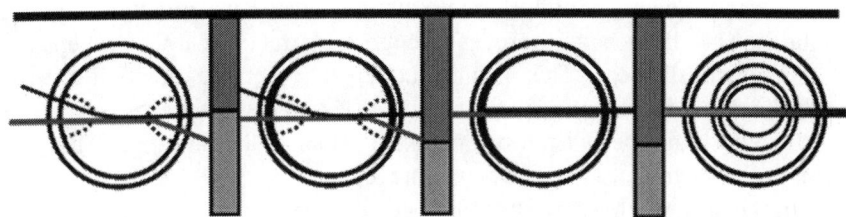

Fig. 1.50 Principle of mechanical splicer

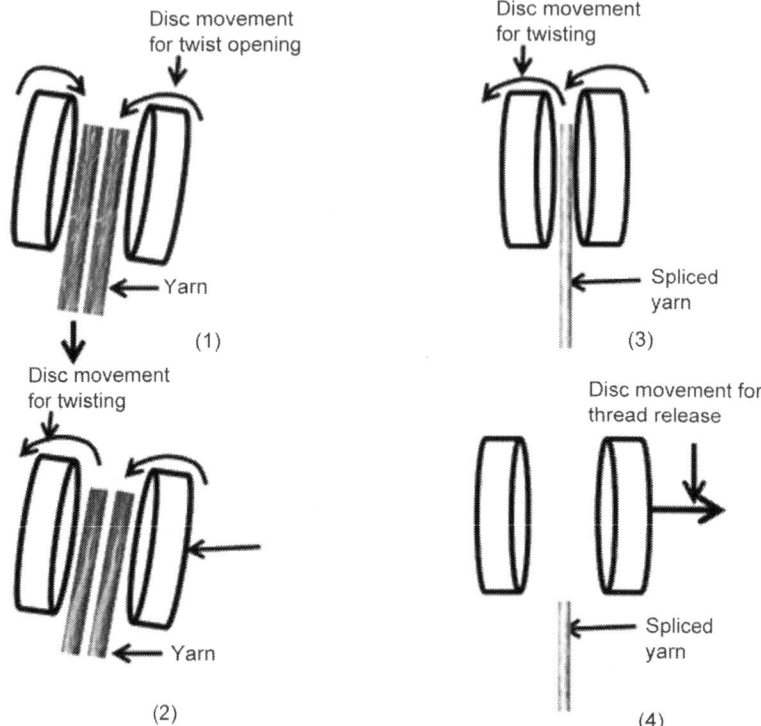

Fig. 1.51 Sequence of operation of discs of mechanical splicer

Sequence of operation of mechanical splicer (Fig. 1.51)
- Discs move in opposite directions to open the yarn twist
- Discs move close to each other to impart twist
- Discs move in opposite directions to splice and impart twist
- Discs open to exit the yarn after splicing

In the first phase, the two pieces of broken yarn are placed between the disks along the diameter of the circle. Rotating one disk in a direction opposite to the other would result in torques of opposing directions at the two ends of the imaginary diameter. This means that depending on the direction of twist in the yarn and the directions of rotation of the two disks, the two ends of the yarn would either be twisted or untwisted. Thus in the first phase, the two pieces of yarn are untwisted and straightened.

In the second phase, the prepared central part is moved in and the tails are pulled out. In the next phase, the yarns are matched and re-twisted. Subsequently, the disks are opened and the splice extracted.

This splicer works on a purely mechanical principle with the advantage of controlling and guiding the yarn in all the phases. The untwisting–re-twisting action of the yarn is done between two interfaced disks with self-compensating properties, so that they can adapt to each yarn diameter. There are three settings that need to be done: untwisting, re-twisting, and straightening. Substances like wax have no effect on the splicing process. Hassen and Sakli (2005) concluded that the yarn linear density has no influence on the properties of splice. Therefore, mechanical splicing may be used for a wide range of yarns. Untwisting and re-twisting are the most influential parameters of the mechanical properties of mechanically spliced yarns. To achieve high splicing quality, an adequate opening of the yarn ends and a convenient re-twisting are required to achieve a successful mechanical splice.

Thermosplicing
This splicing technique is useful to splice the wool, wool-siro, and wool blend yarns. The thermosplicing process is an optimized form of pneumatic splicing in which the air temperature is controlled precisely. Hot air as splicing is a combination of complementary two splicing, blast-enable, and superior quality joints on long staple yarns. The splice air is heated locally inside the splicer to make intensive use of specific properties of thermoplastic yarns. The air track supply (ATS) system has been provided to supply power and compressed air to the hot air splicer. This splicing technique ensures the safe and sustainable fixation of the fibers in spliced portion to achieve high retained splice strength and good visual appearance of spliced joint. Thermosplicer is useful to splice a wide variety of materials like single and plied wool yarns, worsted single and plied yarns, high twisted (crepe) worsted yarns, blend of wool with different thermoplastic fibers, and wool blend with elastane fiber yarns. This type of splicers are able to splice some more sensitive yarns consisting of blends like angora, silk or goat fibers. MESDAN® has developed Hot Jointair 4983B, a hot air splicer.

Injection splicer
The use of the injection splicer is recommended for splicing single and plied yarns of vegetable fibers, for e.g., coarse compact yarns, coarse single and plied cotton yarns, OE-rotor yarns, and linen yarns. The standard splicing process is optimized by the addition of a small quantity of water to the splicing air. The dosing rates are entered at the informator adapted to the yarn. The exact water dosage is injected at each winding unit. The result is spliced joints with higher strength and smoother appearance.

Elastosplicer
The elastosplicer is particularly useful for the splicing of elastic core yarns. Its specific features, including special braking elements, modified clamping and

cutting lines, optimized control software, ensure safe and smooth processing of these springy yarns. The elastosplicer can also be combined with the injection splicing facility. Strength and visual appearance of the spliced joints are of excellent quality. Besides that they convince by their high resistance against alternating stresses in weaving, and they maintain their characteristic elasticity for knitting.

MESDAN® has developed a versatile range of "splicers" in both manual and automatic versions, capable of joining yarns which may differ in fiber nature, spinning method, count, and torsion (e.g., continuous filament, single and twisted spun yarns, open-end, compact, and elastomeric varieties).

Splicers which mingle the two yarn ends using compressed air only are particularly suitable to join wool and synthetic yarns and are marketed as "jointair".Splicers which use water in addition to compressed air are most suitable to join hygroscopic yarns such as cotton and linen, these are marketed as "aquasplicer".

Both jointair and aquasplicer are available in the automatic version for the direct installation on new winding machines, and in a hand-operated version for installation, together with the air track supply (ATS) system on all textile machines such as assembly winders, twisting machines, manual winders, warping machines, and on the creels for knitting together with weft creels in the weaving mills for tail-to-top transfer operations

Our line of splicers and knotters stands out for the exclusive MESDAN® technology, with top quality materials and latest production techniques ensuring a great reliability of performances independently by operator's ability.

Splicers are classified in two broad classes on the basis of sequence of operation of splicing. The classes are given below.
- Ends-opposed splicing
- Ends-together splicing

1.13.3   Ends-opposed splicing

Ends-opposed splicing is the form of joint (Figs. 1.52 and 1.53), which was used when splicing technology was first developed. It makes the best-looking splices.

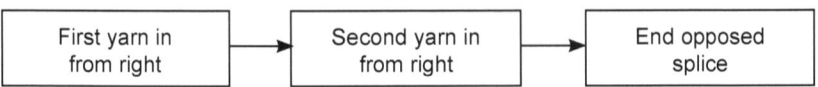

Fig. 1.52 Ends opposed splicing

| First yarn in from right | Second yarn in from left | Ends-opposed splice |

Fig. 1.53 Ends opposed splicing

The two yarns to be joined are placed into the splicer from opposite sides, one from the right hand and the other from left.

- When the splicer operation is activated, the two yarns are joined together by an air blast, rather analogous to that of linking the fingers of two opposing hands.
- Once the waste ends have been trimmed off, the resulting splice has a very flat form.
- The operation takes several seconds to complete, but the results are good.

1.13.4   Ends-together splicing

Ends-together splicing is a very simple form of joint (Fig. 1.54). It was first used to splice yarns, which had proved to be completely resistant to normal ends-opposed splicing.

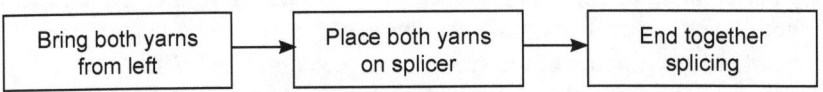

Fig. 1.54 Ends together splicing

- The two yarns to be joined are placed into the splicer from the same side.
- When the splicer operation is activated, the two yarns are joined together by an air blast, rather like the fingers when two hands are clapped together, palm to palm.
- Once the waste ends have been trimmed off, the resulting splice has a distinct "tail" at about 90° to the thread-line.
- The operation is very quick and simple, but the results are only satisfactory for some non-critical applications.

Figures 1.55(a–e) demonstrate the sequence of splicing process on a modern cone-winding machine, employing ends-opposed principle. Splice strength is increased, in particular for yarns of natural vegetable fibers, such as highly twisted cotton yarns, denim and open-end yarns, linen yarns, and plied cotton yarns by the use of injection splicer. Here, a small quantity of distilled water is added to the splicing air. The thermosplicer is normally used for yarns made from animal fibers or blends of these with manmade fibers.

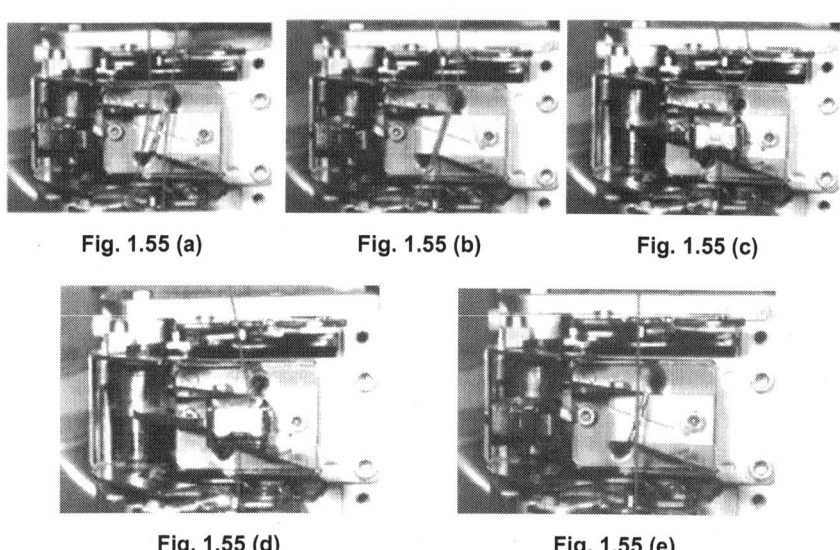

Fig. 1.55 (a)   Fig. 1.55 (b)   Fig. 1.55 (c)

Fig. 1.55 (d)   Fig. 1.55 (e)

For splicing of short staple compact yarns and elastomer-blended yarns of both S and Z directions, manufacturers offer mechanical twinsplicers. The principle of functioning of the system is depicted in Fig. 1.54.

In the first phase, the two pieces of broken yarn are placed between the disks along the diameter of the circle. Rotating one disk in a direction opposite to the other would result in torques of opposing directions at the two ends of the imaginary diameter. This means that depending on the direction of twist in the yarn and the directions of rotation of the two disks, the two ends of the yarn would either be twisted or untwisted. Thus in the first phase, the two pieces of yarn are untwisted and straightened.

In the second phase, the prepared central part is moved in and the tails are pulled out. In the next phase, the yarns are matched and re-twisted. Subsequently, the disks are opened and the splice extracted.

This splicer works on a purely mechanical principle with the advantage of controlling and guiding the yarn in all the phases. The untwisting–re-twisting action of the yarn is done between two interfaced disks with self-compensating properties, so that they can adapt to each yarn diameter. There are three settings that need to be done: untwisting, re-twisting, and straightening. Substances like wax have no effect on the splicing process.

1.13.5   Factors affecting the properties of spliced yarn

The retained splice strength, retained splice elongation, and appearance of splice portion depend on following factors:

Effect of fiber and fiber blends
Fiber properties such as torsional rigidity, breaking twist angle, and coefficient of friction affect the splice strength and appearance of spliced portion. Lower torsional rigidity and higher breaking twist angle permit better fiber intermingling. Higher coefficient of fiber friction (μ) generates higher inter-fiber friction to give better strength to spliced zone. The blend of polyester with cotton, wool, and viscose increases the splice strength.

Effect of number of fibers in yarn cross-section
Higher number of constituent fibers in yarn cross-section, generally in case of coarse yarns, gives better strength in spliced zone.

Effect of yarn twist
Higher amount of twist in parent yarn significantly increases the breaking load and elongation, even at higher pneumatic pressure. Das and Ishtiaque (2005) revealed the fact that after the splice formation there is leaking of twist to the spliced portion, and it takes place from the parent yarns on either side of the joint and hence higher twist level of parent yarn results in more twist leaking. As the splice portion contains less twist as compared to the other part of yarn, so it helps in improving the binding of fibers in this region. So higher splice strength is realized as the yarn twist multiplier (TM) increases. But at the same time, parent yarn strength also increases with TM. So the rate at which strength increases due to leaking of twist to the spliced portion may be higher as compared to the rate at which parent yarn strength increases due to increase in yarn TM.

At high pneumatic pressure, better opening of yarn ends takes place. The splicing of ply and cable yarn are complicated because opposite twists remain present in single and double yarn. Twisted yarn requires relatively higher time for opening of yarn ends.

Effect of spinning methods
Yarn produced with different yarn-spinning systems exhibit different structures and properties. The ring yarn possesses best splicing, but potential of splicing is affected by spinning conditions. The breaking strength of ring-spliced yarn

to a parent yarn is 70–85% for cotton yarn. However, it depends on fiber and yarn properties. The presence of wrapper fiber in rotor spun yarn makes it difficult to untwist and its disordered structure is less ideal for splicing. The breaking strength of spliced zone is 54–71% of parent yarn. In open-end friction spun yarns, the tensile strength remains 80% of parent yarn, but number of splice failure remains higher due to specific yarn structure.

The air jet yarns and core sheath yarns are virtually impossible to splice. Very low tensile strength is obtained with high variation due to inadequate opening of the ends during preparation for splicing.

Effect of opening pressure

Initially, the tensile strength of spliced joint is increased with increase in opening pressure. However, long opening time deteriorates the strength. An increase in pressure up to 5 bar caused release of fiber tufts and fiber loss from yarn ends due to intensive opening in polyester/cotton blend yarn but beyond this pressure, drafting and twisting in opposite direction may also occur.

Das and Ishtiaque (2005) concluded that at constant yarn TM, with the increase in splicing pressure the spliced strength retention (SSR) increases up to a certain level of air pressure and with further increase in air pressure it deteriorates. Due to initial increase in splicing air pressure, the torque increases which facilitates better intermingling of fibers in the overlapped region and binding gets improved. But after reaching saturation level, there is no further improvement in binding of fibers, rather it decreases. This may be due to very high splicing torque which creates turbulence, which has adverse effect on intermingling and binding of fibers, resulting decrease in retailed splice strength (RSS). And also at lower splicing air pressure, as fiber friction value goes up, there is marginal decrement in retained splice strength (RSS); this may be due to ineffective intermingling at low splicing air pressure, and ultimately binding of fiber is not good. At the same time, parent yarn strength increases as fiber friction increases. So the ratio of splice strength to parent yarn strength decreases. But at higher level of air pressure, the trend is just opposite – good intermingling takes place and binding is effective.

Effect of splicing duration

By increasing the splicing duration, the breaking strength and strength retention in spliced portion are increased because of increased cohesive force, resulting from an increased number of wrapping coils in a given spliced portion. However, the splice length should be as small as possible. The splicing duration alone has no conclusive effect on retailed splice elongation (RSE) of spliced yarn. For different types of fibers and yarns, the blast time varies from 0.5 to 1.8 s.

Effect of splicing length

Strength of spliced joint is increased with increase in splicing length. Splicing length is more pronounced on the load-elongation properties of spliced yarn. Effect of wetting

Wet spliced yarn gives more strength than dry spliced yarn. The splicing is more effective for yarn made of long staple fiber and for coarse yarn. Hassen et al (2007) concluded that the yarn linear density and the air blast duration are the parameters that most affect the mechanical properties of pneumatically wet-spliced cotton/elastane yarns. The yarn linear density, the splice length, and the wet flow duration affect the appearance of the splice. In optimal condition, it is possible to obtain high-quality splice (retain spliced strength >90%). The special structure of cotton/elastane yarns has an influence on RSS.

Effect of splicing chamber

The mode of air supply pressure along with type of prism affect the splicing quality. It is observed that irregular air pressure is advantageous over constant pressure for better and effective intermingling in the splicing chambers.

Effect of yarn count

Webb et al. (2009) explored the relationship between yarn count and splice performance in case of synthetic yarn. He revealed that increasing the knife separation vastly improved the yarn count range a given splicing chamber could handle before failing to create acceptable splices. He concluded that the smaller cross-sectional splicing chambers to splice larger yarn counts, improves the resulting splice appearance.

1.13.6   Assessment of yarn splice quality

Ahmadabad Textile Industry's Research Association (ATIRA) has developed various parameters for judging the spliced yarn quality as given below:
- Splicer Appearance Grades (SAG)
- Retained Splice Strength (RSS)
- Splice Breaking Ratio (SBR)
- Retained Splice Elongation (RSE)

Appearance of a splice yarn is also rated on a numerical scale by developing visual standards. These set of boards contain splices of grades 1–7, according to splice appearance.

Cheng et al. (2005) concluded that the yarn linear density and the length of the yarn tails are the dominant parameters that affect the appearance of the splices. However, the splicer manufacturers mainly focus on the tensile properties of the splices. More work on the splicer or splicing chamber design is needed to improve the appearance of the splices.

Retained Splice Strength (RSS) can be expressed as

$$RSS = \frac{\text{Strength of splice yarn}}{\text{Strength of parent yarn}} \times 100$$

Retained Splice Elongation (RSE) can be expressed as

$$RSE = \frac{\text{Elongation of splice yarn}}{\text{Elongation of parent yarn}} \times 100$$

Splice breaking ratio (SBR) is introduced to characterize a splice for its RSS. The SBR is computed by expressing the number of breaks in splice zone (*splice* ± 10 mm) as a percentage of total tests. If in any particular case, lower is the SBR, higher is the RSS; hence, better is the splicing quality. A splice with 40 SBR can be considered as a good quality splice.

1.14   Yarn clearers

After crossing the splicer, the yarn reaches in clearing zone of yarn winding machine where it gets checked for splice, long defects, and repetitive defects such as thick and thin places, foreign fibers, and vegetative matters. The yarn passing through a clearer at a speed of 1000–1500 m/min is scanned either for its dielectric properties by a sensor working on capacitance principle or for its optical diameter reflectivity by optical sensors.

Yarn clearer is a device that is used to remove the objectionable faults of yarn in order to increase the yarn quality and weaving efficiency.

Classification of yarn clearer

Functionally, yarn clearer can be divided into two broad classes as given below:

- Mechanical type
- Electronic type

Mechanical yarn clearers

This is conventional class of yarn clearers in which simple mechanisms are employed to clear the yarn as much as possible. The efficiency of mechanical-type clearer remains less than electronic yarn clearers. Mechanical yarn clearers can be divided in two parts.

(a) Conventional blunt type

(b) Serrated blade type

Both conventional and serrated blade-type yarn clearers are shown in Fig. 1.56

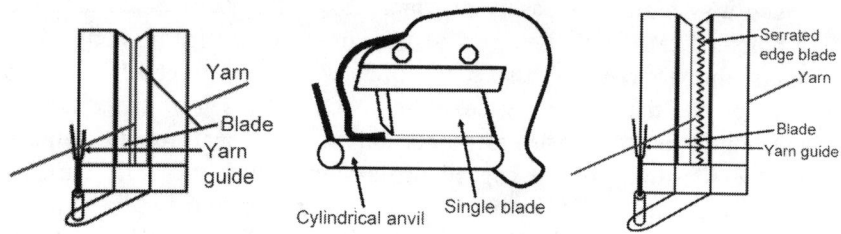

Fig. 1.56 Mechanical yarn clearers

1.14.1   Electronic-type yarn clearers

Capacitance-type electronic yarn clearer
Capacitance is the property of a capacitor to store energy in the form of an electric field between two conducting plates. The dielectric property of the medium in the intervening space between the plates affects the capacitance. All other factors being equal, greater permittivity of the dielectric gives grater capacitance. Glass for instance, with a relative permittivity of 7, has seven times the permittivity of pure vacuum. Consequently, glass will allow for the establishment of electric field flux, seven times stronger than that of vacuum. Capacitance of homogenous dielectrics in a uniform dielectric field is generally proportional to the mass, but is also a function of frequency and for materials like cotton and additional substances (e.g. moisture, and irregular geometry and shape) may affect capacitive measurement results.

Functioning principle of capacitance-type electronic yarn clearer
The yarn is measured in a measuring field, constituted by a set of parallel placed capacitor plates as shown in Fig 1.57. When the yarn passes through this measuring field (between the capacitor plates), an electrical signal is produced which is proportional to the change in mass per unit length of the yarn. This signal is amplified and fed to the evaluation channels of the yarn-clearing installation. The number and type of evaluation channels available are dependent on the sophistication and features of the model of the clearer in use. Each of the channels reacts to the signals for the corresponding type of yarn fault. When the mass per unit length of the yarn exceeds the threshold limit set for the channel, the cutting device of the yarn clearer cut the yarn.

Optical type electronic yarn clearer
Optical sensors employ suitable light sources, reflectors, and receivers for recording the extent of light scattered and absorbed with and without the yarn in the way (Fig. 1.58). The light imaging system of a photo-optical type yarn clearer is shown in Fig. 1.59.

Such a system can measure the thickness of the yarn, as the nature of light reflected can detect presence of foreign substance in the body of the yarn. Assuming circular cross-section of the yarn and a certain degree of homogeneity of the body of twisted fiber assembly, it can be stated that the mass per unit length of yarn is proportional to the square of its diameter. Hence the sensitivity of the capacitance system, which measures variation in mass per unit length, is significantly higher than that of the optical system, which measures the variation in the optical diameter of the yarn. For example, the doubling of yarn mass (100% increase) would result in a diameter increase of 42%, all other factors remaining the same.

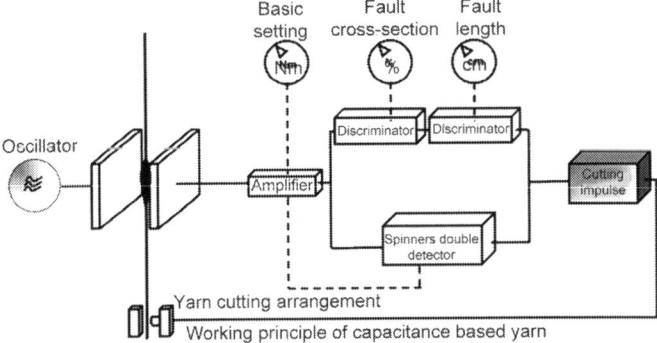

Fig. 1.57 Capacitance type yarn clearer

In spite of its higher sensitivity, the capacitance system suffers from two drawbacks namely, the sensitivity to moisture and inability to isolate the signals for hairs/loops, and foreign matters from that of the core yarn body. In this regard, the optical system has been found to be superior.

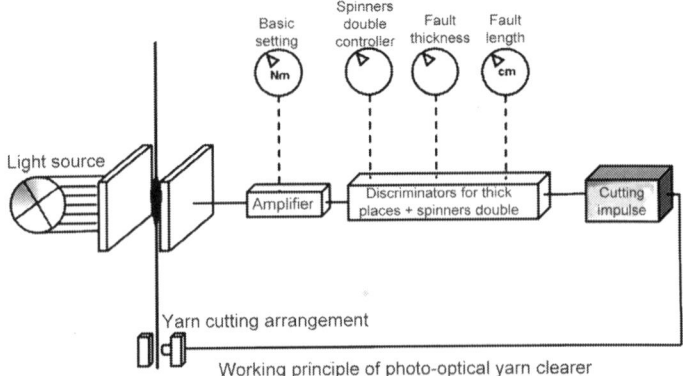

Fig. 1.58 Optical type yarn clearer

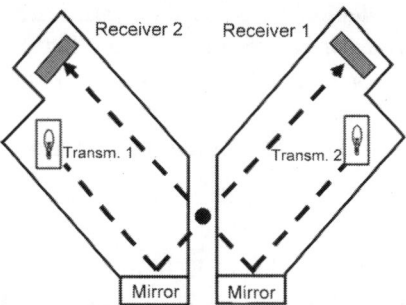

Fig. 1.59 Optical yarn imaging system

Yarn faults can be grouped into two categories: deviation of the mass/ diameter per unit length from a desired value, and deviation of the light reflectance from a desired level. The first type of deviation is caused by slubs, neps, short and long, thick and thin places, and hairs/loops sticking out of the main yarn body while the other type results primarily out of contamination. The desired quality level of the wound yarn would depend on the requirements of the end use and the same can be set at the control panel of the winding machine. The darkened portion in the box refers to the types of undesirable faults. Whenever the threshold is crossed, the cutter comes into action and the faulty segment of yarn is removed, the broken ends are spliced and then the winding is restarted. A spliced joint whose quality has also to confirm to the overall quality requirements thus replaces every objectionable yarn fault. Hence, the clearer is located after the splicer in the path of the yarn, such that a bad splice can be removed and replaced by a proper one.

Yarn clearer setting
The yarn clearer has to be provided with certain basic information in order to obtain the desired results in terms of clearing objectionable faults. Some of this basic information is as under:

Clearing limit
The clearing limit defines the threshold level for the yarn faults, beyond which the cutter is activated to clear the yarn fault. The clearing limit consists of two setting parameters: sensitivity and reference length.
   1. Sensitivity
       Sensitivity determines the activating limit for the cross-sectional size of faulty region of yarn.
   2. Reference length
       This defines the length of the yarn over which the fault cross-section is to be measured. Both the above parameters can be set within a wide

range of limits depending on specific yarn-clearing requirements and knot factor. Here, it is worth mentioning that the "reference length" may be lower or higher than the actual "fault length". For a yarn fault to be cut, the mean value of the yarn fault cross-section has to overstep the set sensitivity for the set reference length.

Comparison between mechanical and electronic clearer

- Electronic clearers are more sensitive than mechanical clearers.
- In case of mechanical clearers, there is abrasion between yarn and clearer parts; but in case of electronic clearers there is no such abrasion.
- Mechanical clearers do not prevent soft slab from escaping through clearer, where as electronic types do not allow passing of any types of faults.
- Mechanical types do not break the thin places and the length of the fault is not considered.
- Mechanical clearers are simple and easy to maintain, while the electronic clearers are costly and require high standard of maintenance.

1.15   Automatic thread stops motion on warp winding machines

Both cone and cheese winding machines operate at very high speed which requires an effective thread stop motion to break the contact of package with rotating winding drum, immediately in the event of an end break. Thread stop motion immediately lifts the particular cone or cheese to protect the continuous rubbing of outermost layers of yarn on package surface against the winding drum, which may cause multiple damages to the outermost yarn layers.

1.15.1   Major problems in absence of thread stop motion

1. After a long run, thread will break on package surface due to excessive rubbing.
2. The ends of broken thread will be embedded and lost in the layers of yarn on package surface, and it will be very difficult for the winder to locate it for knotting or splicing.
3. Identification of broken end drum will be difficult.

1.15.2   Concepts of thread stop motions

Most of the mechanical automatic thread stop motions are worked by the drop of either a drop wire or a drop pin, which obstruct the path or movement of some moving parts. This obstruction later translate or transmit to lift the package from winding drum surface as shown in Fig. 1.60.

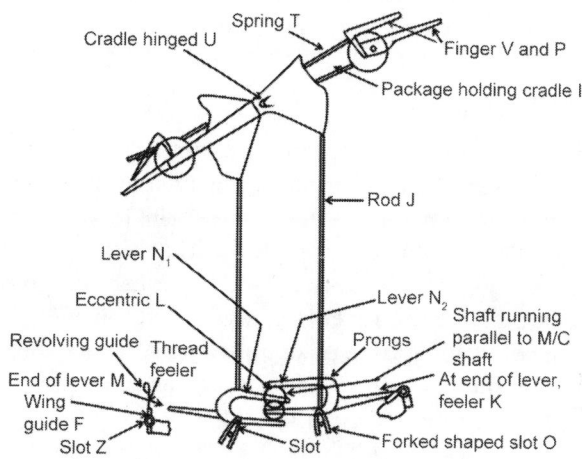

Fig. 1.60 Mechanical thread stops motion

1.15.3   Working of mechanical thread stop motion

As the yarn leaves the tensioning device, it passes through the eye of a drop wire. The purpose of this drop wire is to lift the cheese off the winding drum in the event of breakage of thread passing over the drop wire.

The bobbin cradle is attached with a rod which is fulcrumed in the fork-shaped slot at bottom side. Two levers N_1 and N_2 are found at bottom part of it and work as a feeler because drop pin is attached at the end of lever as shown in Fig. 1.60. The other end of the lever is in the form of a fork. An eccentric revolves between the prongs of fork of lever fitted on a shaft, revolves parallel to the machine shaft. As the eccentric revolves, it causes the feeler to make up-and-down movement along the arc of a circle. The presence of yarn in drop wire held the handle of fork M in upright position but as the thread breaks, the drop pin loses its support, and fork handle and drop pin both fall. The fork takes a horizontal position and consequently obstructing the free movement of the feeler K. When the end N of the eccentric is in its upward position, it lifts the fork end of the lever N_2 up, with the feeler end pressing on M which acts as its fulcrum, the slot Z allowing this to take place. In response of this, fork end M rises and lifts the rod J and consequently lifts the package cradle and in consequence of that package gets lifted up from winding drum surface.

The yarn from drop wire passes through a revolving guide F, which gives a reciprocating movement to the yarn. A pair of light plate segment mounted on a horizontal shaft comprises the wing guide. The segments are inclined at a certain angle and each is cut at one of its ends in such a way that a slot, having a shape like the one on a quick traverse drum winder, is formed for guiding the yarn.

1.16   Waxing

Waxing is required for yarns meant for weft knitting and similar applications in which the package of wound yarn is directly converted to fabric, and the yarn coefficient of friction plays a major role in the fabric formation process. The yarn-waxing unit should logically be the last element in the yarn path, as only the cleaned yarn needs to be waxed to the required degree before being wound on to the package. However on few winding machines, the waxing unit is located even before the splicer. A controlled pressure is applied on the block of wax and the same is made to bear against the moving yarn while being slowly rotated against the direction of yarn flow for ensuring even pick-up (Fig. 1.61).

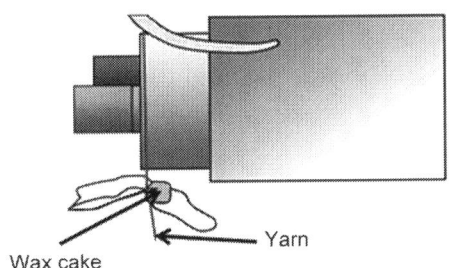

Wax cake   ←——— Yarn

Fig. 1.61 Waxing of yarn surface

The wax has to be removed later on from the yarn and hence should be emulsifiable in a normal scour bath employing common detergents. Usually, the waxes used for yarns have a melting point varying between 48 and 62°C. Closer the ambient temperature of the winding section to the melting point of the wax, greater would be the percentage wax picked up by the yarn. The coefficient of friction of yarn does not however keep on falling with increasing amount of wax picked up. Indeed, as depicted in the Fig. 1.62, after an initial drop (zone A), the value of coefficient of friction remains steady over a range of wax pick up (zone B), followed by a continuous rise (zone C). The wax percentage at which the minimum value of friction coefficient is reached depends on the type of yarn, type of wax, the angle of wrap at the yarn–wax contact point, the yarn tension as also the winding speed. For instance, a 100%

increase in winding speed (from 700 to 1400 m/min) may cause a 50% rise in wax pick up (from 1.2% to 1.8%).

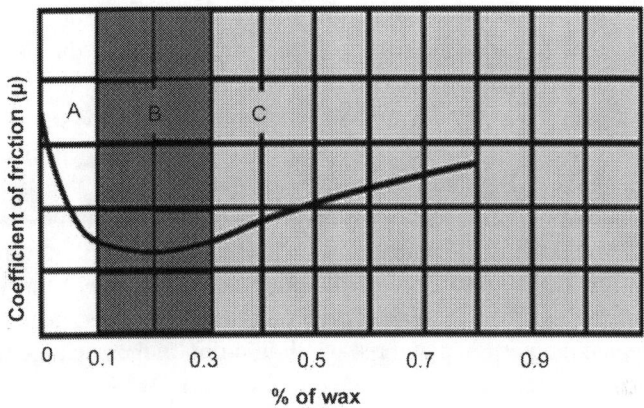

Fig. 1.62 Effect of wax content on coefficient of friction of yarn

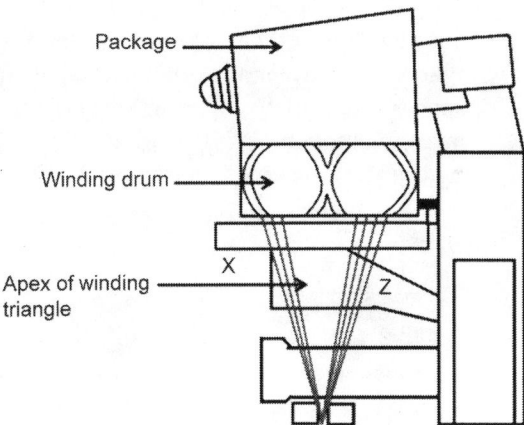

Fig. 1.63 Apex of winding triangle

The yarn ultimately passes through the yarn trap, or finger of thread stop motion shown in Fig. 1.59. It retains the lower yarn end in the event of a thread break. It also functions as a thread guide, which determines the apex of the winding triangle, shown in Fig. 1.63. The base of this triangle is formed by the projection on the vertical (say the XZ) plane of the locus of the yarn element entering the drum groove. The projection of the actual path of yarn from the yarn trap to the drum groove on the three mutually perpendicular planes would exhibit shifts. The shift is governed by the helical groove of the drum and shift of cam in case of precision winding, which forces the yarn to execute a to-and-fro motion. As the path length along the two sides of the triangle is the longest and that along the altitude the shortest, one would expect this shift to cause a periodic tension fluctuation in the yarn segment above the yarn trap. The locus of the yarn end entering the drum groove would also trace a path along the XY and YZ planes, caused by the periodic change in depth of the drum groove. These shifts would have added effects on the tension fluctuation. Beyond the drum groove entry point, the yarn segment follows a narrow curved channel within the groove before landing on the body of the package. Barring the path along the altitude of the winding triangle, this channel is inclined opposite to the direction of movement of the yarn along the base of the winding triangle. The projection of the yarn path on the XZ plane between the yarn trap and the winding point on the package would thus exhibit kinks of varying degrees. Hence for most of its journey from the yarn trap to the package, the yarn path exhibits a highly complex configuration. It can therefore be stated that in spite of the tension-controlling systems before the yarn trap, a less ordered condition prevails at the final winding stage, in so far as the nature of yarn path is concerned. Indeed the variation in wall angles of the groove at various locations of the drum, as also the variations in grove depth, coupled with the wrap between the drum and the yarn, varying in the range of 90–120°, appear to be solely aimed at keeping the yarn within the groove during the traversing motion.

The yarn faces a very harsh treatment during its stay in drum groove and that definitely possesses a remarkable problem in the rotary traverse system of drum-winding machines.

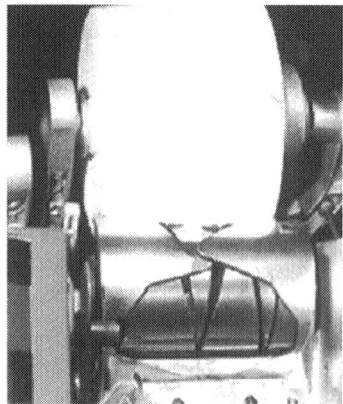

Fig. 1.64 Package holding by two adaptors of cradle

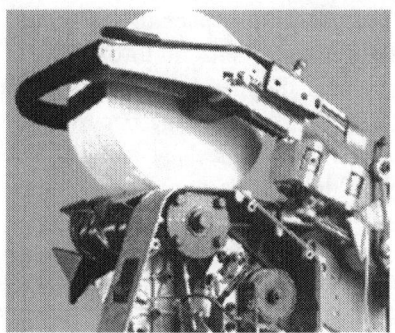

Fig. 1.65 Package holding by right hand adaptor

The package is held between two adaptors of a cradle, as depicted in Fig. 1.64. The right hand adaptor can be shifted laterally against spring pressure to allow the package sleeve to be gripped firmly as shown in Fig. 1.65. A sensor in the adaptor is used to record the number of package rotations. The modern winding machines are capable to suppress the yarn length and package diameter variation within range of 2–1%, respectively. The cradle can be adjusted to deliver packages of conicity varying from 0 to 5°57′. Usually, lower package conicity is associated with shorter traverse. The conicity can on requirement be increased even up to 11°. The right hand package adaptor is provided with a pneumatic braking system for bringing the package to a quick halt in the event of a yarn brake or bobbin run out, as depicted in Fig. 1.66.

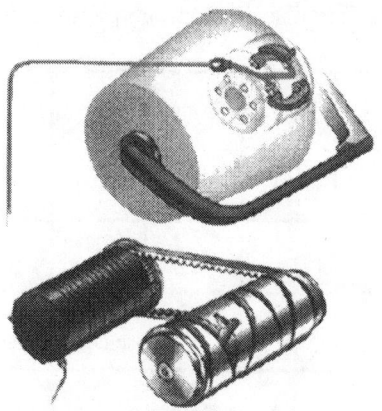

Fig. 1.66 Package run out

The cradle pressure on the package can be adjusted for producing soft, medium density, and hard packages. Increasing cradle pressure due to increasing cradle weight of the package is compensated, keeping the operational cradle pressure nearly constant throughout the package build. The cradle linkage is oil-cylinder damped, eliminating vibrations. The center of the grooved drum and that of the package occupy the same radial distance from the fulcrum of the cradle (Figs. 1.64 and 1.65) so that line of contact between the package and the drum remains nearly undisturbed during entire build of the package.

Each grooved drum of a modern drum winding machine is driven directly by a servomotor. The drum might be mounted directly on the shaft of the motor or be driven by toothed belt (Fig. 1.65). Drum lap guard and drum lap brush are provided at the back of the drum to prevent development of lappers on the drum. Deflector plates cover the drum at the front side, either partly or fully (Fig. 1.64). After a completed yarn joining cycle, the deflector plates guide the yarn towards the center of the drum, so that the jointed yarn slides safely and securely into the guide groove of the drum. Moreover during bobbin run out or yarn break, the lower contours of the deflector plates guide the free upper yarn end towards the drum center. The servomotors driven drums are programmed to vary the winding speed in such a manner that the unwinding tension in yarn being withdrawn from the supply package remains within desirable tolerance limits. As shown in Fig. 1.67 (also shown in Fig. 1.10), the unwinding tension rises steadily as the unwinding point shifts from the tip of the bobbin towards other end. The servomotor can be programmed to gradually reduce the unwinding speed (Fig. 1.68) in a desirable manner so that the tension fluctuation is suppressed to the required extent. This feature supplements the controlled disc tensioning system (Fig. 1.32).

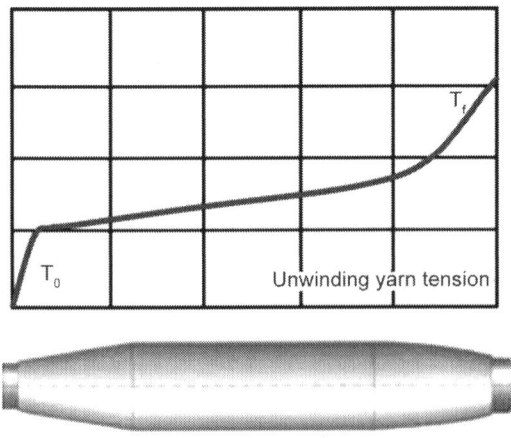

Fig. 1.67 Yarn unwinding tension

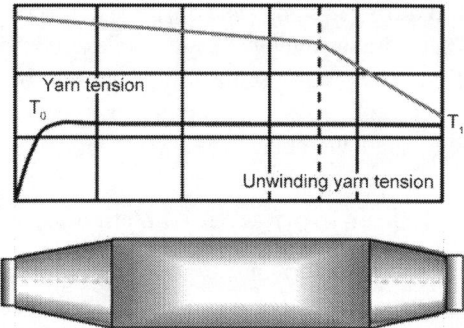

Fig. 1.68 Yarn unwinding tension and drum speed

The angle of spirality of the helical grooves of a drum, i.e. the angle that the groove makes with the drum axis, may exhibit a continuous change from one edge to the other. The spiral angle near the drum end on which the nose of the cone rests would be lower than the angle near the drum end supporting the base of the cone, and this difference would increase with package conicity. This leads to the gradual increase in surface area of a conical package and hence the necessity of slowing down the traverse speed, as traverse moves (yarn winding point) from the nose to the base of the cone for maintaining constant yarn length per unit surface area. The angle of spirality also exhibits a gradual rise in the neighboring two edges of the drum. This strategy of drum design offers a smooth transition of the clockwise spiral (from extreme end of traverse in one direction) into the anticlockwise (into other direction of traverse) one and vice-versa during the change in the direction of traverse. Such high angles would lead to considerable rise in the amount of yarn wound near the nose and near the base of the cone that consequent a harder base and nose of the package than rest of the package body. To overcome this problem, a traverse motion is integrated with drum, which ensures a slow lateral shift of the drum relative to the package. A modern winding machine is equipped with number of auxiliaries to assure the high speed winding of quality packages consistently. Few of the auxiliaries are discussed below:

- The central computing station where all important input variables are fed in via the touch screen, and which controls the working of different elements such as the various servomotors in a synchronized manner.
- The conveyers of full and empty ring bobbins, as also of the full and empty cones and sleeves.
- The full package doffing system that patrols on tracks located above the winding units. The doffer winds a nose tail on the completed

package and doffs the same. The doffer also places a new cone/sleeve from the magazine into the package cradle and winds a transfer tail.
* The traveling cleaning system which patrols the entire frame and sucks away loose fibers/yarns from various locations and deposits in a centrally located chamber.

1.17   Latest developments in drum winding

Higher production, lower power consumption, better quality, and a suitable package for subsequent processing are the major driving forces of latest developments in drum winding.

Apart from a lot of modern developments in drum winding machine, the basic objectives of the winding process remain unchanged. These developments are directly focused to achieve:
* Efficient and judicious fault removal
* Higher productivity
* Minimum damage to the yarn
* Uniformly wound package

1.17.1   Features of a modern surface-driven cone winding machine

Typical winding units of a modern cone/cheese winding machine are shown in Figs. 1.69 and 1.70, respectively. The supply package, namely the ring bobbin, is housed in a magazine in Fig. 1.69, whereas in the other case (Fig. 1.70), it is mounted on a peg.

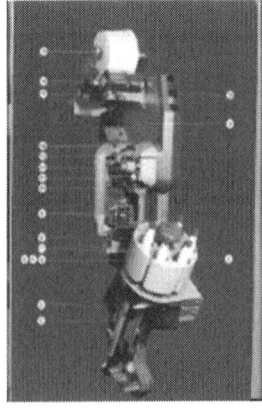   

Fig. 1.69 Modern cheese winder   **Fig. 1.70** Modern cone winder

Replenishment of the magazine is a manual process, whereas complete automation in feeding is achieved in the other system, which involves a conveyor system that removes the peg carrying the exhausted bobbin by one carrying a full bobbin.

An important component of the yarn supply unit is the booster/balloon controller. This device adjusts itself continuously and maintains a constant distance from the package nose. The balloon is thus kept under control, limiting the fluctuation in the unwinding tension.

The central unit of the winding unit carries:

- Tensioner
- Splicer
- Electronic yarn clearer and
- Waxing unit.

The tensioner of a modern machine senses the yarn tension continuously and employing a feedback loop maintains the same at a constant level. Figures 1.69 and 1.70 illustrates such a system in which the sensor is visible at the top part and the disc tensioners are seen in the center. The discs are driven slowly by a motor against the running yarn thus preventing deposition of dirt particles, wear or even cutting of the yarn into the disc surface.

The pressure between the discs is varied as a function of the instantaneous tension sensed by the system. The progressive rise in unwinding tension resulting out of a constant disc pressure at a winding speed of 1250 m/min is observed in Fig. 1.71, while a constant although a slightly higher unwinding tension under a progressively falling disc pressure at a much higher winding speed of 1450 m/min is revealed in Fig. 1.71. Thus, the booster of the supply unit in association with the yarn tensioner can ensure that the yarn is fed to the winding head at a reasonably constant tension.

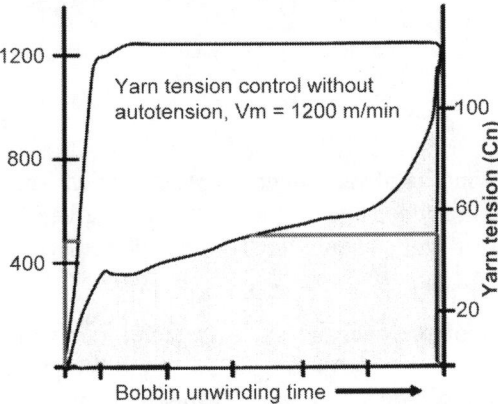

Fig. 1.71 Progressive rise in unwinding tension

1.17.2   Mechanism to avoid ribbon formation or patterning

The grooved drum, the major part of drum winding unit, is dedicated to traverse and wind the yarn to the package. Grooved drum winders are very much prone to ribboning.

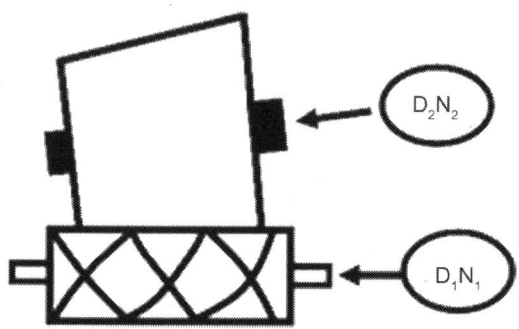

Fig. 1.72 Grooved drum and package

1.17.3 Condition of ribbon formation

Suppose a grooved drum revolves at rpm N_1 and makes x revolutions per double traverse of the effective winding point. The package revolves at N_2 rpm. The diameter of the drum and package are D_1 and D_2 respectively as shown in Fig. 1.72. Neglecting slippage between the package and the drum, the surface speed of both drum and package should be equal, i.e.

$$\pi D_1 N_1 = \pi D_2 N_2$$

$$\frac{D_1}{D_2} = \frac{N_2}{N_1}$$

$$\text{Or} \quad N_2 = \frac{D_1}{D_2} \times N_1$$

Continuous winding of yarn on package leads to change different phases, as different layers are added one by one on package surface. This phase change occurs as each layer is added, and it can be represented as follows:

$$\phi = \pi D_2 = m\lambda$$

The diameter of the package D_2 is variable, λ is yarn traverse, which is fixed and m is an integer. As the package build progress, ϕ changes and at a particular time ϕ reduces to zero. At this particular time moment, yarn from one layer is wound exactly upon the previous layer, and the yarn will be wrapped up in ribbon form till the diameter increases enough to give ϕ a significant value again as shown in Fig. 1.73. The situation of ribbon formation will come again when m reaches successively larger integers. The ribboning or patterning occurs when ϕ reaches zero, as well as a fraction of $\dfrac{\phi}{pd}$ i.e., $\dfrac{\phi}{pd}, \dfrac{3}{2}, \dfrac{5}{2}$ and so on.

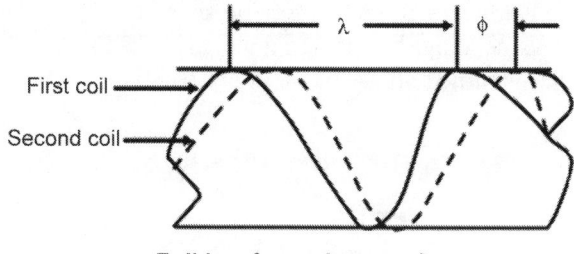

Build-up for package surface

Fig. 1.73 Yarn traverse and phase difference

Ribboning or patterning leads to:
- Heterogeneous yarn dyeing due to uneven package density in case of yarn dyeing
- Poor unwinding and slough-off of the package
- End-breakage in warping and weaving
- Needle damage in knitting

Above points are enough to understand the importance of anti-patterning or ribbon-breaking devices as an integrated part of modern automatic winding machines. Different leading winding machine manufacturers have invented various devices to prevent ribbon formation and facilitate production of ribbon-free package. Few selected anti-patterning mechanisms of different winding machine manufacturers are as follows:

Double pitch drum
The multiple-grooves anti-patterning mechanism was introduced in 1980s on "Pac21" model of Murata automatic cone winding. This automatic cone winder monitors the critical ribbon causing parameters, and shifts the yarn from one groove to another groove on the same drum to avoid ribboning. At the time of patterning, the jumping mechanism switch over the yarn from winding at 2 coils to 2.5 coils wind. Manufacturers of Pac21 claimed that this winder is able to process all types of yarns, count, and winding packages.

Schlafhorst Propack® system

This anti-patterning mechanism was introduced by Schlafhorst/Saurer in Autoconer 338 winding machine. The patterning occurs when the number of coils laid on the package per double traverse is a whole number. Winding unit computer of the Propack® system constantly determines the ratio between the drum and the package rotational speed. As the critical speed ratio is about to come which produces pattern, the Propack® system reduces the pressure on the cradle by a pre-set amount. In consequence of that, the package runs at slower speed below the critical patterning speed till the package diameter is adjusted to a value above the pattern zone, then the Propack® cradle anti-patterning device is turned off as shown in Fig. 1.74.

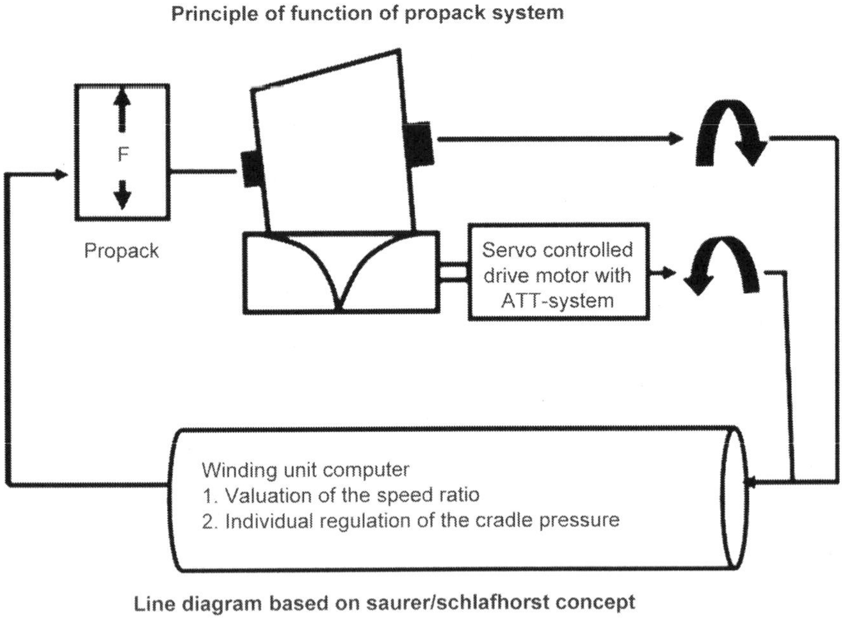

Principle of function of propack system

Line diagram based on saurer/schlafhorst concept

Fig. 1.74 Schlafhorst Propack® system

1.17.4   Computer aided package (CAP®) building

CAP® building is an electronic solution to produce ribbon-free package. This device was first opted by Savio on its automatic winder "Orion". CAP building system controls the ratio of package diameter to drum diameter by means of a computer on each winding head. In this system, a servo motor is used for automatically intervening in the drive ratio between package and drum to control the yarn deposition on package surface as shown in Fig. 1.75. Savio claims that perfect packages without patterns can be manufactured with CAP® on Orion winders.

Consequently, no unwinding speed limits are imposed in the subsequent processes of warping, knitting, and dyeing. Few of the other devices which make Orion automatic winder, a "state of the art" machine, are inspector control, tensor, balloon-breaker, booster (optional), etc. The most prominent feature of Orion is the positioning of the electronic clearer as the last device on the yarn path. Any long or more defects are detected by the clearer and eliminated by being sucked from the package.

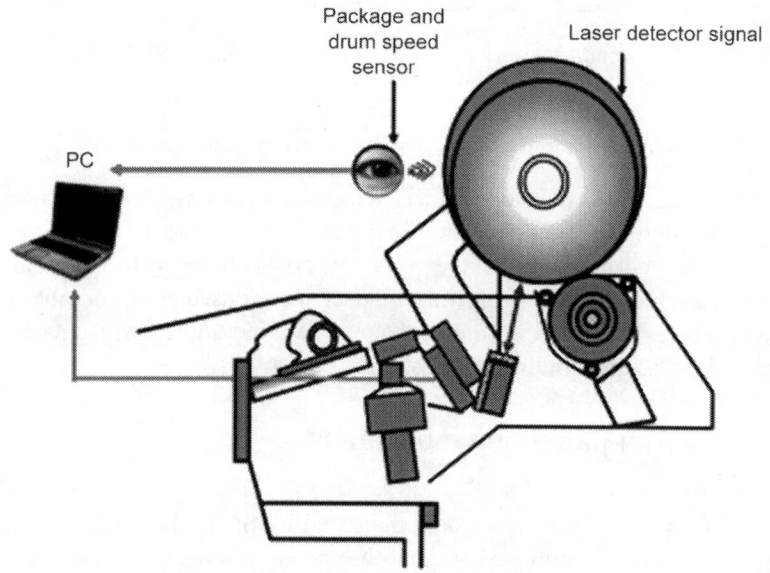

Fig. 1.75 Computer aided package (CAP®) building

1.17.5   Semi-conductor devices (TRIAC)

Semi-conductor device (TRIAC) is used by Murata's MachConer automatic winder as anti-patterning device. Anti-patterning is achieved by means of dial setting for on and off durations regarding drum motor. Thus, the anti-patterning is achieved by the adjustment of speed change cycle. As the drum moves, the package is constantly being accelerated and decelerated, the ratios of drum and package speeds never reach critical value to produce ribboning as shown in Fig. 1.76.

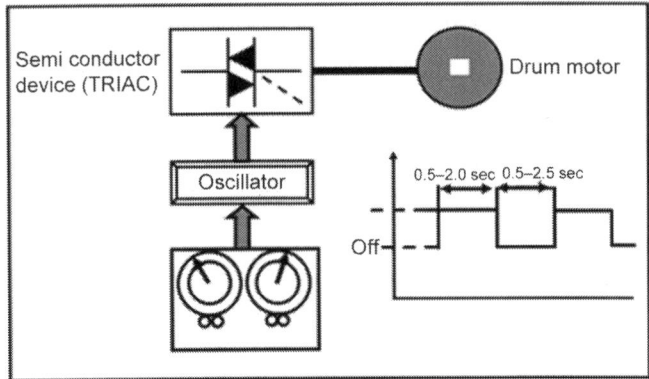

Fig. 1.76 Semi-conductor device (TRIAC) [*Source:* Murata Tech technical leaflet]

1.17.6   Counter rotating blades with precision winding

On PSM-51 of SSM precision winder, two counter rotating blades are used to avoid patterning. The function of these blades is to traverse the yarn wrapped on the cylindrical package. Blades rotate at constant speed to maintain the constant traverse ratio from empty to full package. Consequently, no ribboning or patterning is produced on the package. The wind angle shrinks narrower progressively as the winding progresses.

1.17.7   Anti-patterning modulation

"On-off" type electronic modulation is a regular device incorporated in all models of Orion automatic winders marketed by Savio. Drum and package speeds are constantly adjusted, so that the speed ratio does not reach critical values, and patterns on the package are minimized.

1.17.8 Developments for hairiness reduction

Winding process is known to increase yarn hairiness due to rubbing against various surfaces of different parts of the machine. Various machine manufacturers are coming up with the solutions to overcome this hairiness problem. Murata Tech Muratec Japan has taken lead in hairiness reduction technique at yarn winding machine.

1.17.8   Hairiness reduction by Perla-A and Perla-D

Perla-A and Perla-D model of automatic winder manufactured by Murata Autoconer, Japan, offered hairiness reducing system with the tension manager.

These winders enable the packages to be wound with reduced hairiness. These yarn-hairiness-reducing devices are equipped just below the tensioning system. An air vortex is used in the yarn path to wrap the protruding fibers on yarn surface. Reduction of only long hairs is ensured by manufacturers to maintain the characteristic feel of spun yarns. Other benefits claimed by the Murata Group are less stoppages of loom due to reduced shedding caused by entanglement of warp yarn, and less fly generation at subsequent processes with 30% less size pick-up. Comparison between Perl-A and Perl-D model of Murata Tech is shown in Table 1.3.

Table 1.3 Comparison between Perl-A and Perl-D model of Murata Tech.

Type	Perla-A	Perla-D
Yarn count	Ne 10s to Ne 100s	Ne 10s to Ne 100s
Method	Air nozzle	Friction discs
Energy	15 NL to 20 NL/min/head	40 W/head
Yarn speed	Up to 1800 m/min	Up to 1200 m/min
Control	Compressed air pressure	Rotational control by Servo-motor

Tension management
Uniform tension is essential to maintain homogeneity in package density. Modern automatic winders are able to maintain the uniform yarn tension right from start to end of the package at lower tension level. Various approaches offered by different manufacturers for better tension management are as follows:

Bal-con and tension manager by Murata
Balloon formation during yarn unwinding from ring frame bobbin is one of the major causes of tension build-up in running yarn on winding machine. Hence, a balloon control device can be proved more successful to maintain the tension uniformity. . Bal-con stands for balloon control and equipped on Murata automatic winder. Bal-Con tension manager of Murata is shown in Fig. 1.77. This Bal-con has facility to lower supply package when it is desired to maintain even winding tension from start to end of the winding. The combination of Bal-con with solenoid gate tensioner forms the tension manager. Bon-con maintains a constant balloon height with varying unwinding point throughout the unwinding of the copes. Hence maintains a constant tension. The gate-type tensioner offers required pressure to the yarn along with movement of Bal-con and hence tension management becomes possible even at higher speed. The yarn tension is controlled by opening and closing of the gate-type tensioner . The yarn passes exactly through the middle

of the gate-type tensioner that gives proper gripping on the yarn at higher speed; the yarn tension is maintained stably at a desired level. The function of tension manager is to sense the position of the yarn on the ring bobbins and send signals to computer. Computer controls the yarn tension by sending the command to solenoid of gate tensioner to manipulate the pressure on the running yarn accordingly. Gate tensor adjusts the pressure on the running yarn in such a way that winding tension remains constant. In case of slow speed winding process after each splicing, or winding from a full ring frame bobbin, pressure is set at high level with the help of computer-aided program in the computer to keep the winding at constant tension. When winding takes place from bottom of the ring bobbin, height between balloon control ring and unwinding point at ring bobbin increases; pressure is set low to maintain the winding tension at the same level as previous portion of the bobbin.

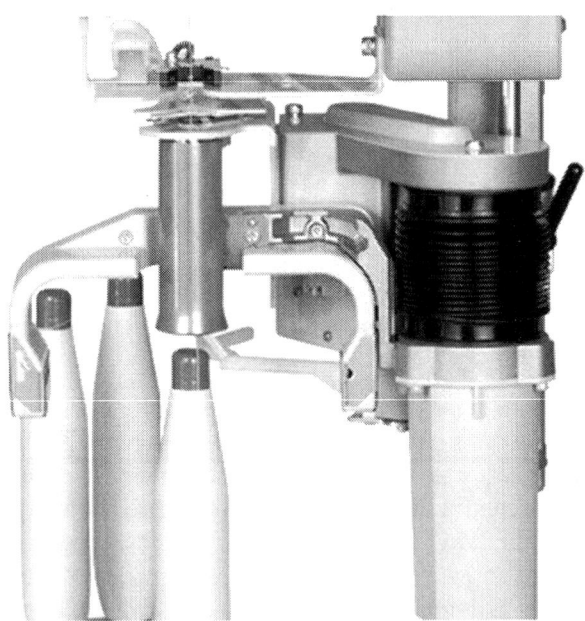

Fig. 1.77 Bal-con technique by Murata Tech

The tension manager system controls tension fluctuation at winding point after yarn joining and at around the end of winding, based on balloon control detected supply data and by the pressure induced tension governed by the gate-type tensioner on each spindle. The gate-type tensioner maintains tension in running yarn by manipulating the pressure electronically, so that setting becomes easy without replacement of any parts.

The tension manager ensures uniform winding tension at high speed winding, from empty to fully wound package without reducing the winding speed. The manufacturer of Bal-con claimed that lowering down of the balloon controller over the supply package, depending upon the height of unwinding point, offers advantages of reducing the possibility of sloughing-off and suppressed hairiness. Unwinding from supply packages becomes easy because Bal-con minimizes the contact surface between the yarn separated from the bobbin and the yarn layer on the bobbin. The Bal-con mechanism brings the possibility of sloughing-off very negligible in comparison of conventional balloon-breaker mechanism.

1.18   Contribution of different drum winding manufacturers

1.18.1   Contribution of Savio

The winding segment of textile industry has been always directed by the needs of world textile market. Savio, an Italian company, accepted this challenge and provided an appropriate solution to the market forces and introduced the POLAR automatic winder. POLAR automatic winder is available in a number of versions:

POLAR M/L – This is available with manual bobbin feeding and manual/automatic package doffing.

POLAR E/I – This is available with a stand-alone machine or for direct linkage with spinning machine in many configurations.

POLAR/I DLS (Direct link system) – This is available with uninterrupted movement of the bobbin from the ring frame to the winding head. This winding machine is equipped with a newly configured yarn-end finder point, targeted to reduce the splicing time.

The main features of the new model POLAR/I DLS (direct link system) includes the uninterrupted movement of the bobbin from the ring spinning frame to the winder. The machine is equipped with a newly configured yarn-end finder point, targeted to enhance the machine efficiency. This device can be extended to second and third point in order to accommodate the production of the longest ring frames. The winding path also has been modified in POLAR/I DLS. The position of the supply bobbin is vertical to winding drum with two spare bobbins in each position.

Suction system has been modified also with improved aerodynamic performance blades and a direct-drive motor. The direct-drive motor is controlled by the machine's computer via an inverter that automatically tunes its speed according to the actual process parameters and settings.

Computer-aided tension (CAT) system

Savio introduced CAT system that enables the continuous tension measurement of yarn to facilitate online controls of the tension of the yarn. The winding tension is continuously measured by a tensioner device, based on piezoelectric system, which interacts with the yarn tensioner via the machine's computer and modifies the load on the yarn in order to keep the winding tension constant during whole bobbin unwinding. The CAT system does not require continuous contact between the yarn and the sensor. The tensioner device is located just before the package, which measures the actual winding tension and acts as an anti-wrap system.

Savio bobbin identification system

This system is able to detect the ring frame spindle which produces faulty bobbins. Additionally, this system is equipped with a centralized station which automatically removes the faulty yarn from the detected faulty bobbin. A Savio winding machine, dedicated to process the coarse count yarns, is equipped with a dust removal unit, located at the level of balloon-breaker, in order to catch all fibrils and dust generated during bobbin unwinding. The necessary vacuum is created by an independent fan serving for two sections. The dust is discharged in a collecting box with a separate filter.

Savio computer-aided package (CAP) system

This system is able to control the amount of yarn deposit on package surface to eliminate ballooning. This system is dedicated to process wool and other delicate yarns. A step motor modifies the drum and package position relatively to ensure the desired "gain" between two successive coils on package to suppress the possibility of ballooning.

Savio air splicing systems

Savio contributed to develop better air splicing systems to improve retained splice strength and splice appearance also.

KZ906L automatic air splicer

KZ906L (690L) automatic air splicer is equipped to the Savio automatic winder "Orion". This splicer enables the processing of a wide variety of yarns. All splicer related settings are set at the computer integrated with "Orion" winder to reduce the setting up time at the time of lot change. Splicer related data of any spliced yarn can be recalled when required. The KZ906L splicer is precision built and manufactured with the highest grade of material available and is sealed against penetration of dust and fugitive fibers, the result of which is high performance with minimal maintenance.

Savio's computer aided metering (CAM) system

The CAM system ensures the accuracy of yarn length on each package. The request for a reduction in yarn remnants in downstream processes requires that the packages produced contain an extremely precise yarn length. The resulting savings are particularly welcome in case of value-added yarns such as wool, cashmere, and silk. The combination of the laser detector with the package and drum sensors is managed by the machine's PC. This measuring system is totally independent of the structure, shape, evenness, and hairiness of yarn.

1.18.2   Contribution of Muratec

Murata Machinery (Kyoto, JAPAN) has launched a latest winding machine of 21st century in October 2001 with a product name 21C Process coner. 21C Process coner machine is equipped to provide higher flexibility, higher production, and consistent quality with reduced cost of operations. The cost of operation is reduced by suppressing the energy consumption, labor, and maintenance. 21C Process coner is capable to reduce hard waste and being user friendly.

Few unique features of 21C process coner

21C Process coner winding system is equipped with a combination of tension manager (Bal-con gate tensor) and straight yarn path (SYP) and to achieve an optimum performance on all 60 spindles by opting following accessories:

- *Tension manager:* A combination of gate tensor and Bal-con is attached on 21C Process coner. This is most suitable method of controlling yarn tension variation during high speed winding.
- *Straight yarn path (SYP):* The yarn path from the supply cope to the waxing device is exactly straight. This enables the yarn to be wound at a much higher speed without damaging its quality.
- *PAC 21 drum (combination of traverse ratio 4 and 5):* 21C Process coner is designed to provide the flexibility of winding in terms of yarn linear density, package type, and yarn kind. The traverse changing system of this machine at critical points helps to wind a ribbon-free package for high speed unwinding in the next process.
- *Visual on-demand system (VOS):* VOS is a color monitoring system with advanced information management system that is able to show data for individual winding head in both figures as well as in graphical form. This enables an easy control of machine operation on all 60 winding heads.

- *Nine creel magazine:* This nine creel magazine increases the feeding capacity of each winding head by 60%. Consequently, it improves the operator efficiency which guides to higher productivity.
- *Power management:* In combination with VOS which constantly monitors and controls the negative pressure required by the machine, power management system reduces the power consumption.
- *High speed auto doffer (no. 21D):* High speed auto doffer system is able to maintain 60 m/min traveling speed and 9 s doffing cycle, which increases machine productivity and efficiency. It has an additional program to detect and segregate package quality by monitoring the abnormal number of EYC cuts and informing operator by a special stop-motion named as unit alarm function.
- *PLC 21 (perfect length counter):* It is a new yarn length measuring system which has been developed with the latest sensor technology and contributes to make a perfect length package to reduce the hard waste generation at warping stage.
- *Unit display system:* Individual winding head equipped with a dedicated display system with different color light to signal about the action.
- *Hard waste reducer:* Double arm action of splicing unit enables a better grip of yarn on the ring-cope during splicing. This decreases the length of yarn which is sucked as hard waste.
- *Perla-A and Perla-D (hairiness reducing device):* These devices are also used by Muratec to improve the yarn quality by reducing hairiness. These devices are also used in Savio's winding machine Orion to suppress yarn hairiness.
- *Water splicer:* Muratec is an established name of air-splicing technology. However keeping in mind the growing demand of the market, Muratec has further developed the air splicing system for different usage, 3 tier air-splicers for specialty yarn, and the water splicer for 100% cotton double yarn.

With these developments, 21C Process coner is ready to process compact yarn, cotton lycra yarn, denim yarn, denim with lycra yarn, multi count yarns, cotton and P/V double yarns, poly/wool with lycra, 100% wool siro yarn, etc. 21C Process coner is a very advanced machine to serve the continuous changing needs of the textile industry to manufacture yarn package of superior quality.

Figure 1.78 represents a single winding head in which all essential parts and mechanisms are shown. Rotary circular magazine is used to feed the ring-frame bobbins continuously one after one, as winding process continues and ring bobbins get exhausted due to yarn unwinding from it. Electromagnetic tensioner is used to regulate the yarn tension in order to produce a package of homogeneous density. Pneumatic splicer is used to produce knot-free yarn. Electronic yarn clearer is used to regulate the yarn imperfection in running yarn. Auto yarn tensioner device is used after clearer to tune tension variation minutely. An arrangement of wax application is integrated to modify the yarn surface for better smoothening. Finally, yarn reaches on package after passing over grooved drum.

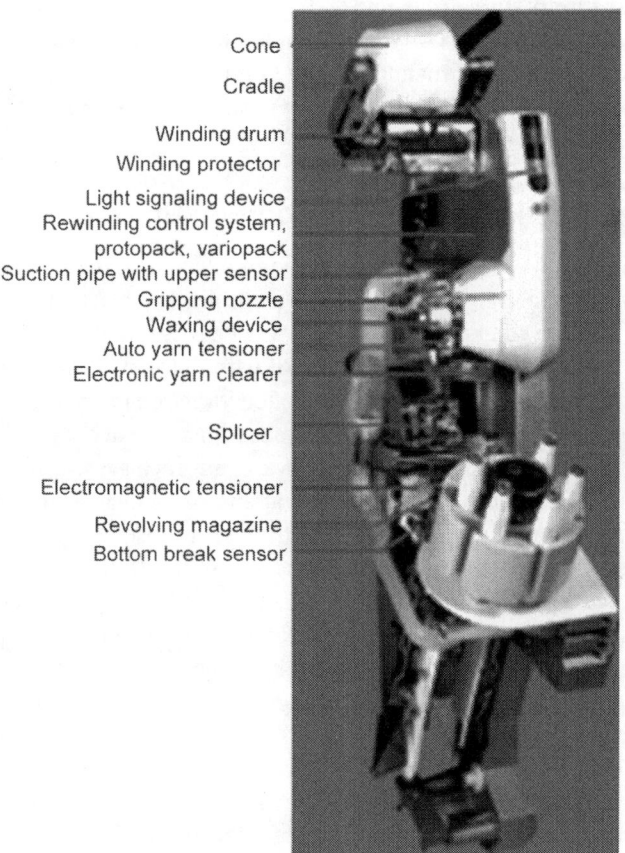

Cone
Cradle
Winding drum
Winding protector
Light signaling device
Rewinding control system, protopack, variopack
Suction pipe with upper sensor
Gripping nozzle
Waxing device
Auto yarn tensioner
Electronic yarn clearer

Splicer

Electromagnetic tensioner
Revolving magazine
Bottom break sensor

Fig. 1.78 Single winding head

In conclusion, it may be stated that the foregoing account of the anti-patterning devices incorporated in their automatic winding machines by the major winding machinery manufacturers to produce ribbon-free packages is by no means exhaustive. Pakistan's textile industry has been installing maximum number of Murata Mach coner, followed by Savio espero and Orion winders. However, the latest trends are in favor of import, installation and commissioning of the Schlafhorst Autoconer 338, as part of the BMR programs of the spinning mills.

Possible uses of a cone
A cone produced on a winding machine acts as a basic unit, which can be used
   1. To produce pirns of weft for shuttle weaving machine or spindles of yarn for braiding machine.
   2. To produce beams of warp yarns for weaving, warp knitting or some nonwoven fabric manufacturing systems.
   3. Directly on shuttle-less weaving machine as weft or on weft knitting machine as supply package.

1.19   Research and development in warp winding

Porat et al. (1999) did computer simulation to design a tension compensation system to prevent long-term tension variations (such as those caused by increased package diameter) with increasing package diameter. This situation generally arises in case of constant rate of yarn delivery system (for example open end machine). Compensation in tension variation is performed by means of feed-back tension monitoring device that adjusts the driving roller speed to maintain the constant tension in running yarn, but such system is very difficult to incorporate in cone winding because at any particular driving roller, yarn take-up speed or requirement varies from base to nose part of the cone. The roller speed and yarn take-up speed equals at point of drive (POD). Porat et al. (1999) did a computer simulation of a cone winding system with constant rate of yarn delivery to locate the POD and prove that POD is dependent on yarn tension and on the frictional force between driving roller and package. This computer simulation derived an agreement by which the new tension compensating system eliminates the effect of increasing cone diameter on yarn tension up to a major extent. The basic concept used in this simulation was to monitor yarn tension and to correct any deviation from its desired value by adjusting the speed of the driving roller that rotates the receiving package. Finally, Porat et al. (1999) concluded that the available software can quite easily be developed to take account of additional winding parameters such as increased cone size during winding, inertia of the cone, effect of cone size on friction, pattern-breaking, etc., and further work along these lines seems well worthwhile.

Lang et al. (2004) studied the mechanism of staple yarn hairiness during the winding process. Generally, two types of actions take place on yarn during the winding process – one pulling the fiber ends and the other to provide the resistance. The hairiness of staple yarns formed by trailing protruding fiber ends is analyzed first in terms of the interactions of the fiber ends, the tension disk, and the grooved drum. The possibility of fiber pulled out depends on the ratio of forces of pull and resistance. Lang et al. (2004) explored the problem of yarn hairiness changes due to contact with the tension disk and grooved drum during the winding process. They focused on mostly occurring hairiness of the protruding fiber and more specifically on trailing hairiness. The effects of winding on yarn hairiness are determined by yarn structural characteristics such as the twist factor T_y and the embedded fiber length S, and the winding conditions including the pressure W, exerted on the yarn and the winding tension T. They concluded: first, the effect of tension disk to establish a relationship of whether such fiber pulling out actually takes place is established by a ratio K of the total pulling out force, to the gripping force exerted on the fiber during the winding process. The value of K is also used to study the influence of the grooved drum and cheese on yarn hairiness. The fiber pulling force depends on frictional coefficients of inter-fiber, coefficient of friction between fiber and tension disk, coefficient of friction between fiber and grooved drum, and compression force between grooved drum and the cheese/cone.

1.20   References

- Barella, A. (1957). Yarn Hairiness: The Influence of Twist, *J. Textile Inst.* **48**, p. 268.

- Brunnschweiler, D., and Mohammadain, I. S. (1959). Experimental Studies of Variation in Yarn Tension During Unwinding from Ring Frame Packages, *J. Text. Inst. Proceedings,* pp. 74–113.

- Bhattacharyya, S., Mondaĺ, S., and Pal, S. (2003). Measurement of Yarn Tension and Its Online Monitoring, *Indian Journal of Fibre & Textile Research,* **28**(4), p. 418.

- Benerjee, P. K. (2006). Classnotes IIT Delhi, Department of Textile Technology.

- Cheng, K. P. S., and Lam, H. L. I. (2005). "The Appearance of Pneumatic Spliced Short Staple Ring Spun Yarns" *RJTA* **5**(1).

- Das, A., Ishtiaque, S. M., and Parida, J. R. (2005). "Effect of Fiber Friction, Yarn Twist, and Splicing Air Pressure on Yarn Splicing Performance", *Fibers and Polymers,* **6**(1), pp. 72–78.

- Eiichi, K., and Tetsuya, S. (1964). A Study on Tension Device, *Journal of the Textile Machinery Society of Japan,* **10**(3), p. 139.

- Foster, R. (1959). Tension Variations Occurring During the Unwinding of Cops and Pirns, *Journal of the Textile Institute,* **50** p. 7.

- Goswami, B. C. (2002). Nonlinear Dynamic of High Speed Transport for Staple Yarns, National Textile Center Annual Report.

- Hassen, M. B., Jaouachi, B., Sahnoun, M., and Sakli, F. (2007). "Mechanical Properties and Appearance of Wet-spliced Cotton/Elastane Yarns", *Journal of the Textile Institute,* **98**(1), pp. 119–123.

- Issa, K., and Grütz, R. (2005). *AUTEX Research Journal,* **5**(1), pp. 1–19.

- Jung, J. H., Kang, T. J., and Youn, J. R. (2004). Effect of Bending Rigidity on the Capstan Equation, *Textile Text. Res. J* **74**(12), p. 1085.

- Latzke, P. M. (1979). Yarn Tension Measurements in the Textile Industry, *Melliand Textilberichte,* **60**, p. 207.

- Moser, B. (1989). Basics of Friction in Textile Technology, *Melliand Textilberichte,* **70** (1), p. 33.

- Paul, A. (1982). Length Measurement in the Winding Department, *Int. Textile Bull Weaving,* p. 2.

- Rust, J. P. (1992). Yarn Hairiness and the Process of Winding, *J. Textile Res.* **62**, p. 685.

- Schmenk, B., Wulfhorst, B., Gries, T., and Schedukat, N. (2003). New Developments for Online Quality Control in High Speed False-twist Texturing, *Man-made Fiber Yearbook.*

- Ukey, P. H., Kane, C. D., and Shah, H. R. (2010). "Studies on tension measurement in winding".

- Webb, C. J., Waters, G. T., Liu, G. P., and Thomas, C. (2009). "The Influence of Yarn Count on the Splicing of Simple Continuous Filament Synthetic Yarns", J. Text. Res., **79**(3), pp. 195–204.

Pirn winding

2.1 Introduction

The quality and nature of weft package is very important to decide the weaving efficiency of a loom. Production of an appropriate weft package for loom shuttle is a matter of great importance. In case of shuttle less weaving, a cone can be directly used as weft package but in case of shuttle looms, an appropriate package is required which must be suitable to accommodate with shuttle design and loom requirements. For this purpose, yarn from cone or cheese rewind and wind on a small package called pirn, and this process is known as pirn winding or quill winding (Conrad et al., 1953). Due to the fast emergence of shuttle-less weaving, the pirn winding departments are disappearing from textile industry day by day; but being a basic process, pirn winding is the part of this book.

The pirn winding machine is used to produce a suitable package on wooden pirns after rewinding the yarn from cones, cheese, bleached, and dyed hanks, etc. The supply packages, either cones or hanks are mounted in the creel section of the machine. The yarn from supply package passes around a tension rod and then reaches on pirn through a traverse guide. The traverse guide consists of a duckbill, a bend glass or ceramic rod, to support the yarn on its bend. The traverse guide makes an upward and downward movement (to and fro in case of horizontal spindle-pirn winding machine) to wind the yarn on entire length of pirn according to stroke of traverse bar (Lord and Mohamed, 2005). The traverse guides are fitted on a horizontal rod which consequently receives the motion to get the appropriate shape of pirn with thread. There is a pinion on driving shaft of the machine which drives a mangle wheel through a train of wheels. On the shaft of mangle wheel, there is a cam on which a lever is fulcrumed at a point of framing as shown in Fig. 2.1. The other end of the lever is connected with a rod which extends the entire length of frame. A number of traverse guides, one for each spindle is fitted at regular intervals on the rod. When the small pinion gears with the teeth of mangle wheel on the outside, it drives the mangle-wheel in one direction. When the pinion comes just on the last teeth near the opening, it slides in and gears with the teeth of mangle-wheel to revolve it in opposite direction, resulting in the traverse guide to move up and down carrying the yarn with it. At the start, the winder places the wooden pirn on the spindle with its tips pointed downwards. The flat end of the spindle is then fitted into the rectangular slot of the wharve spindle, after the winder has fastened the yarn onto the pirn. The yarn starts now to wind on pirn surface and traverse guide movement decides the shape of full pirn. Hence shape of the pirn controls the motion of the traverse guide, because traverse guide moves slowly when it moves from base to tip side of the pirn; but the traverse guide moves fast in the journey from tip to base direction of the pirn. In this way, binding of coils takes place during backward movement of the traverse bar. This also causes more yarn to be laid at the upper part of its traverse. A metal cone G which resolves by the frictional contact of the pirn will lift the pirn slightly in each successive traverse. This mechanism is used to form a pirn with cylindrical body with conical top. When pirn gets fully wind, it will be lifted to such extent that the flat end of the pirn spindle comes out from the slot of the wharve spindle, and put the pirn out of action, and new empty pirn comes in operating position.

2.2   Types of pirn winding machines

Our textile industry is using different types of pirn winding machines, and they differ in many ways and can be classified into six different classes (SenGupta, 1970).
1. Primitive age pirn winding machine with vertical spindle
2. Horizontal spindle pirn winding machine with horizontal spindle with reciprocating traverse
3. Horizontal spindle machine with traverse by reciprocating horizontal spindles
4. Pirn winding machine with differential spindle speed drive
5. Spindle-less pirn winding machine
6. Automatic pirn winding machine

2.3   Primitive ordinary pirn winding machine

Primitive-age pirn winding machines are very simple in construction. A cone or hank is used to supply the yarn for quill or pin pirn winding. A thread guide and guide rod are used to give a path to the running yarn. Required tension is introduced by means of a simple tension rod which works on multiplicative tensioner principle. Finally, yarn reaches on pirn after passing through a thread guide. This thread guide makes an up-and-down reciprocating motion for vertical spindle primitive-age pirn winding machine. Lever system is used to keep the pirn in exact winding position. Pirn is mounted on a vertical spindle that receives movement from a driving drum by means of tape-motion transmission.

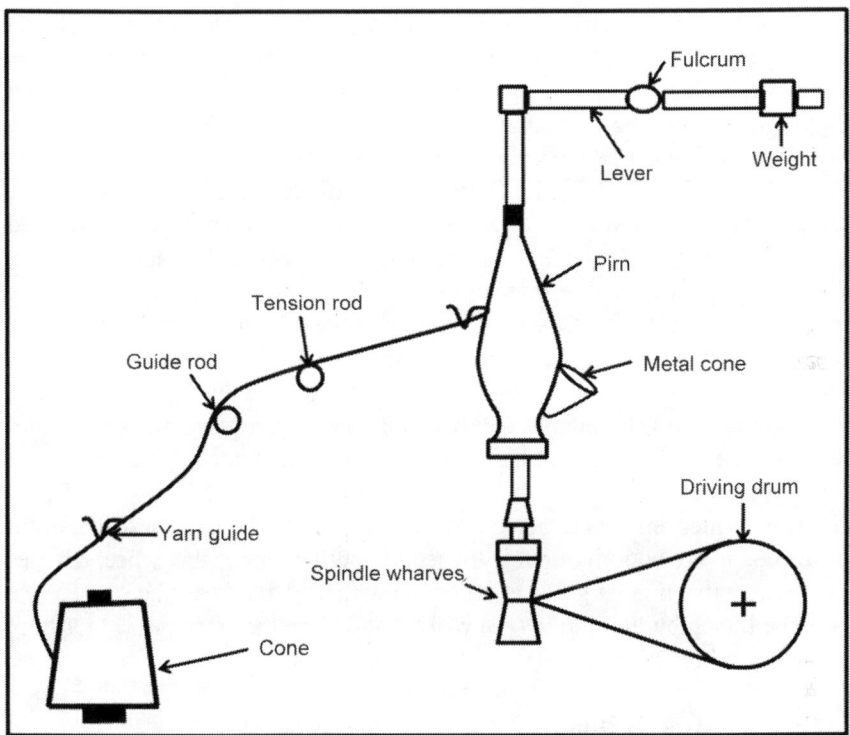

Fig. 2.1 Passage of material on simple pirn winding machine

2.3.1   Bunch motion

Bunch or reserve is defined as a small length of yarn wound on to the pirn near its butt end. Generally, it consists of a length of yarn equal to about three times the width of cloth or length for 3 pick insertion. It is well known that there is always a time lag between the instant when the weft feeler feels the necessary for a pirn change and the instant when the actual transfer operation is carried out by auto pirn or shuttle change mechanism. The bunch ensures the supply weft thread during this time lag to avoid crack formation or loom stoppage (Adanur, 2001).

The automation of shuttle loom is directed to the replacement of pirn or shuttle after exhaustion of yarn on pirn surface without stopping the loom.

The weft winding machines are equipped with a building mechanism to wind the yarn on pirn surface in a definite fashion. The feeler of automatic pirn or shuttle change mechanism feels the amount of yarn on pirn surface by real surface-to-surface contact to send a mechanical message to actuate the pirn replacing mechanism. All pirn winding machines have some kind of package-building mechanism which forms a bunch of yarn on pirn near to its base. Generally, 6–7 yards of yarn is wound in a bunch form (at nearly zero wind angles) for pertaining replacing mechanism to operate. The length of yarn wound in bunch can be altered by a ratchet wheel which is adjusted in such a way that the bunch formed may be of the same length on all spindles. The ratchet wheel works in association of two screws by which it can be adjusted in relation to the ratchet wheel. The relationship between the cam and ratchet wheel controls the number of strokes that the traverse bar makes before the bunch builder guide is released. For each stroke which the traverse bar and yarn guides make, the ratchet wheel is turned one tooth. One end of the bunch builder guide (g2) is held against the face of the cam under the ratchet wheel by spring (g3). As the ratchet wheel and cam continue to rotate, the builder guide will be held in position, guiding the yarn on the bunch being formed on the empty pirn.

The builder remains in contact with surface (g4) and cam (g5). This cam is attached to the bottom of ratchet wheel (g). When the ratchet wheel reaches such a point that builder guide is in contact with the straight surface (g6) of the cam, the bunch builder guide (g2) will be pulled by the spring (g3) to a stop, as shown in Fig. 2.2.

2.3.2   Bunch failure

The bunch formation consists of restructuring the traverse of yarn on pirn base at the starting of winding for a short time span. This is performed automatically by bunch-forming mechanism. Although the pirn winder is equipped with bunch-formation mechanism, a bunch is not formed on butt of pirn. Few of the possible causes are as follows:

1. Bunch-formation mechanism not working properly
2. Improper control over the yarn at the start of winding
3. Incidence of a end break during bunch formation

Bunch failures due to first two causes require immediate maintenance for the machine. In case of failure by third cause, a bunch-less pirn is bound to pass to further processes. Bunch-less pirns are undesirable from view point of automatic loom practice.

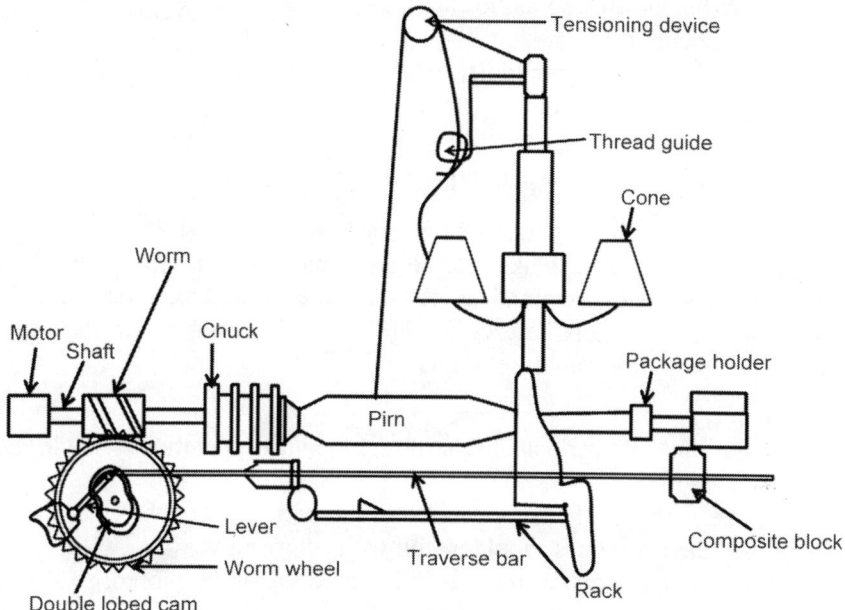

Fig. 2.2 Pirn winding machine by Messrs Thomas Halt Ltd., UK

2.3.3   Locating the bunch

For building a bunch, a bobbin is pushed on the spindle and the yarn is threaded through the guides and thread guide. The thread guide is pressed back extremely on traverse bar to the base of the quill until a click sound is heard, which is caused by the automatic re-setting of the cam and ratchet wheel. The yarn is looped over thread guide, and spindle is started to start the weft winding in bunch form.

Adjustment of ratchet wheel
The ratchet wheel is set to regulate the length of yarn in the bunch. For cotton yarn, the ratchet wheel is usually set to pick about seven teeth.

2.4   High speed pirn winding

High speed winding machines may be broadly divided into two classes. One is spindle less and other is provided with a spindle. High speed pirn winding machines have the following features:
1. Mechanism must be able to rotate the pirn from 2000 to 5000 rpm.
2. Coil locking device, particularly at nose part of the pirn

3. Broken thread stop motion
4. Full bobbin stop motion
5. Pirn holder or pirn spindle, as the case may be in high speed machine
6. Efficient traverse motion

2.4.1   Features of automatic high speed pirn winder

Modern high speed pirn winder is equipped with a supply lift box. As the starting lever is pushed, brake release takes place to start the motor. The supply lift box is lifted automatically by the cables till it strikes the lever; the box turns over and keeps reserve (dumps) the entire bobbin into the chute from where they slide into the bobbin hopper.

Starting operation
Before starting the various heads, the following operations should be performed:

- The cones or other supply packages should be put on the creel.
- All the magazines should be full with empty pirns.
- One empty bobbin in the bobbin chute should be placed, so that it may fall into the feeder.
- Wrap the end of yarn on an empty bobbin, and place the bobbin between bobbin stop and counter stop.
- The running of thread should be observed.
- Start the automatic head by means of starting lever.
- Automatic winders will continue one after another.

2.5   Haccoba pirn winding

In Haccoba pirn winder, there is a magazine containing four compartments for four winding heads. It is situated at one side of the machine, and filled with empty pirns. During winding, the pirns are held in appropriate position by spring-loaded top-end supporters.

- On Haccoba machine, the pirn diameter is controlled by means of graduated setting of yarn traverse mechanism which is more reliable in comparison of disc- or shaper-cup-type diameter controlling devices.
- To accommodate different length of pirns, four end supporters are supplied with machine which can be mounted on a movable head.
- The machine is filled with a stop motion which comes in contact with thread and when it breaks, all four pirns come to a stop. The broken threads are repaired by a quick release motion which releases a grip on pirn, sufficiently to allow them to be revolved by head.

- It is equipped with an adjustable bunch motion to wind a bunch of yarn of varying length on the pirn base.
- The machine is adoptable for automatic pirning of all kind of threads, cotton, wool, linen, and rayon, etc. This machine is capable of winding a large range of counts.
- The machine is capable to achieve spindle speed of 600 rpm.
- The machine is equipped with full pirn stop-motion, auto doffing of full pirn, and auto pirn change in winding position and loose thread cutting scissors.

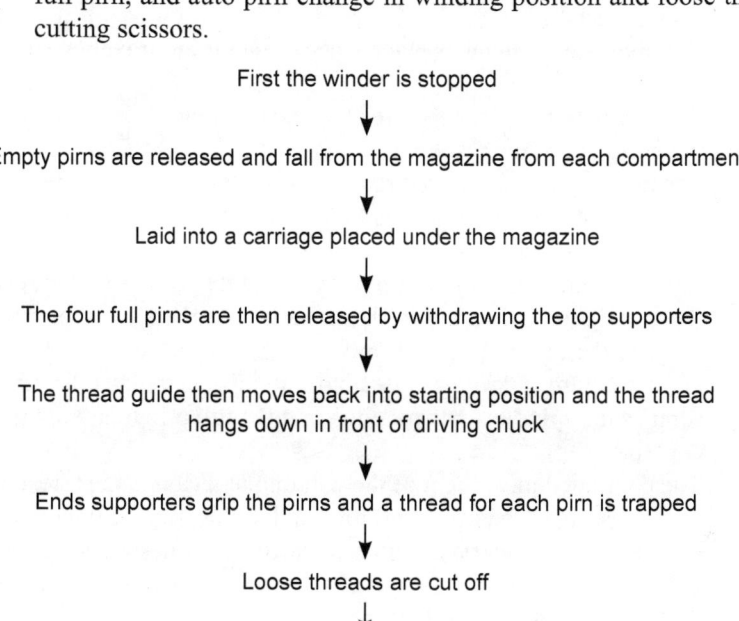

First the winder is stopped

↓

Empty pirns are released and fall from the magazine from each compartment

↓

Laid into a carriage placed under the magazine

↓

The four full pirns are then released by withdrawing the top supporters

↓

The thread guide then moves back into starting position and the thread hangs down in front of driving chuck

↓

Ends supporters grip the pirns and a thread for each pirn is trapped

↓

Loose threads are cut off

↓

Winding starts

2.5.1   Reserve bunch for automatic looms

This machine is capable to wind a reserve bunch of yarn of suitable length. The bunch itself may be wound with the long traverse or short traverse.

2.5.2   Control of tension and pirn diameter

Control of pirn density is provided by a range of standard tensioning devices (disc or gate type). The diameter of pirn is controlled by a wheel according to the setting of which the movement is given to the yarn traverse. The pirn diameter obtained with any given setting is dependent upon the diameter of the yarn. As the yarn changes, control is reset to match with new conditions.

2.6   Automatic pirn winder by Hopper

The main features of this pirn winding machine are:

- The winding unit travels round the machine, and it has only one automatic head to a complete machine of about 100 spindles.
- At a time, four different type of yarns (differ in color or count) can be processed on this machine.
- Various winding units automatically discharge the full pirn into their respective bins.
- Winding operation takes place when the units are traveling round the machine.
- The rate of travel is so regulated that the winding of pirns is completed in one round of the machine.
- The empty pirns are placed randomly into a large hopper attached to the automatic head. From this hopper, empty pirns are automatically placed in winding position to replace the full pirns. In this position the thread leading from the full pirn in the chute lies across the rear end (base end), and when the fresh pirn is inserted and clamped between centers, the thread is also firmly held between the base of the pirn and the centre. A knife situated close to this centre cuts the thread as soon as the spindle starts rotating, and the full pirn then falls into the container.
- The continued movement of the winding unit causes the thread guide to assume its correct position for winding the bunch. Winding starts by means of an electrical switch, a bunch of any desired length is then formed. Stationary cams control the movement of guide during the building of bunch.

2.7   Quality control in pirn winding process

There are few important check points to assure the quality of pirns on any high speed pirn winding machine.

- *Yarn tension:* It must be set appropriately. It can be generated by using dead weights, by using spring loaded tensioners or by using gate-type tensioning devices. For same count, the yarn tension must be same on all spindles to wind the package of uniform density and compactness.
- *Bunch motion:* The dial must be set properly to ensure correct number of turns at the base of the pirn. Bunch length varies from 2 to 15 m.
- *Tension release:* Tension disc must be adjusted to 1.5–2.0 mm above, as per requirement.

- *Bobbin build:* The initial point of wind and the package build must be checked regularly.
- *Bobbin diameter:* It must be checked regularly.
- *Positioning of thread knife:* It must be positioned properly to ensure the accurate cutting of thread end.
- *Full bobbin doffer:* Automatic full bobbin sensor and doffer helps to maintain the same length on thread on all pirns.
- *Speed variation control:* Package should be revolved at different speeds when thread wrapping takes place in different portions of the pirn.
- *Conveyor chain on bobbin hopper:* Chain must move smoothly and adhering yarn ends should be removed.

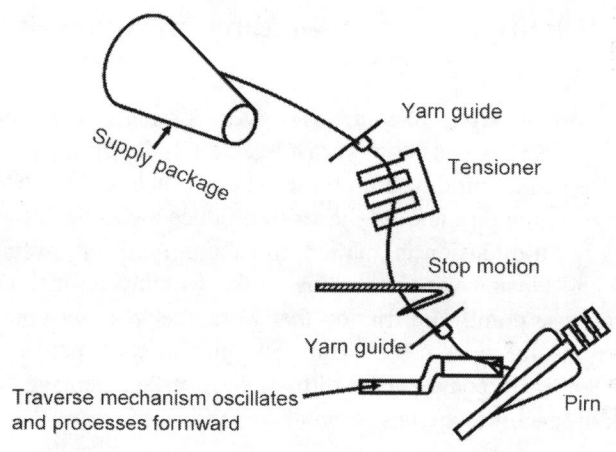

Fig. 2.3 Passage of material on automatic pirn winder

If the yarn is to be used as weft thread in shuttle looms, it must be repackaged on a quill (pirn). The pirn is designed to fit within the shuttle. There are few spinning machines in existence which wind the filling (weft) yarn directly on pirns in place of ring bobbins. The productivity of these ring frames is limited and hence this strategy is not commercialized with economic consideration. To be more flexible, most mills use special winders designed specifically for the purpose of winding pirns (quills).

A schematic representation of a quill winder is given in Fig. 2.3. It is very evident from Fig. 2.3 that a quill winder differs very little from a warp winder, but traverse mechanism is entirely different. In case of warp winding, the traversing mechanism makes a full cycle in traversing the yarn completely back and forth along the package. In quill winding, however, the traverse only covers part of the quill at a time. When one section is built up, the traverse indexes to the next section. This is called building a quill by chase lengths or chasing a quill, and the traverse is called a progressive reciprocating traverse. This method of quill winding is used for the following reasons:

(1) To reduce the tendency of balloon as the yarn is unwound from the quill.
(2) To maintain uniform tension in the filling yarns.
(3) To reduce the possibility of sloughing-off.

If the quills are not to be used immediately after winding, they usually are taken to a room to be conditioned with hot and humid air. This conditioning is done to allow the filling yarn to relax, reducing the twist liveliness of the yarn, and preventing the formation of kinks.

2.8   Primitive age pirn winding machine with vertical spindle

Initially, the weft winding machines were very simple in construction with constant spindle speed and slow yarn traverse (Rude, 2005). The package shape and size can be produced on these types of machines with the help of pressure between the pirn and its shaper to produce a package of appropriate hardness. Very little tension is applied on running yarn on these machines. Yarn speed and tension vary according to the winding on diameter at pirn shoulder and nose point. The friction that works between pirn and the fixed metal shaper results in severe abrasion and high friction between pirn and fixed metal shaper. In consequence of that, yarn surface damage takes place. These types of machines are best suitable to wind linen yarns.

2.9   Major features of N-Tex pirn winding machine

- Compact and sturdy construction. Elegant, economic, easy maintenance, and trouble-free operations
- Rigid quality control, 24 h working before dispatches
- Strict supervision by highly qualified technical staff, backed by 30-year experience.

2.9.1   Functions and uses

Being an automatic pirn winding machine, it is most ideal for winding synthetic polyethylene fiber yarn, cotton (10 counts to fine–superfine), filament yarn, silk, nylon, terry line, terry cot, and woolen yarn.

2.9.2   Operations

- Empty pirn (bobbins) used for winding are arranged in an adjustable magazine, according to pirn size.
- When the dimensions of pirn are adjusted, the rotating pirn is held between the back and front carriers and is also adjusted by moving the slipper box.
- The desired pirn diameter can be adjusted by regular wheel, mounted on left of spindle box. The cones are placed on platform, above the machine. The yarn is unwounded from the cones passing through tension device, stops motion eyes and yarn guides, and winds the pirn.
- Being an automatic machine, it is automatically operated for pirn changing when the required length of yarn has been wound on the pirn. The wounded pirn is replaced by the empty one. Simultaneously, the weft is caught again and cut in time. Moreover, the machine stops automatically in case of breakage of yarn without loss of time.

The basic unit consists of spindles. It can be joined together to obtain 8 spindles. Each unit of 4 and 8 spindles are driven individually. Each unit of 4 spindles is driven individually by ½ HP electrical motor connected to a counter shaft, mounted underneath the spindle box. Eight spindle units are driven by the gears, mounted in spindle box. Oil circulating device is provided for continuous and thoroughly lubricating the specially developed fiber gears for smooth and long life. Other parts are made from highly graded materials to obtain higher efficiency and zero maintenance.

Table 2.1 Technical features of pirn winding machine.

	Minimum	Maximum
Pirn length		
Standard	150 mm	190 mm
First type	190 mm	225 mm
Second type	225 mm	250 mm
Bunch length	2 m	15 m
Spindle speed		
For cotton	3000 rpm	7000 rpm
For silk	3000 rpm	6000 rpm
No. of traverse per lift	6–10	13

2.10   Weft winding defects

2.10.1   Oversized pirns

The oversized pirns will create difficulty during their fitting inside the shuttle. This type of pirn is produced when oscillating thread guide bar is bent and does not make full oscillating motion.

2.10.2   Bobbin with hollow wing in the centre

Due to weak spring which is attached with yarn tensioning device creates pirn with hollow wing in its centre which is a serious pirn winding defect. The major cause of this defect is heterogeneous stress distribution on weft thread during winding on pirn surface from its base to tip

2.11   Priyalaxmi pirn winding machine

The main features of this machine are as follows:
- Easy to operate and maintain
- Individual head is adjustable for different count of yarn and pirns also
- Low noise and low maintenance cost
- Specially designed system for loading the weft empty pirns which ensure the safety, easy, and damage-free pirn loading
- Minimal yarn waste
- Different types of pirns can be used by simple setting
- Highly improved work efficiency ensures higher production
- Work load is less and fewer operators required
- Excellent engineering after sales service
- Placement of cones with proper guided tension ensures 100% equal tension on both spindles.

Table 2.2 Specifications of priyalaxmi pirn winding machine

	2 spindles unit in 1 head
	4 spindles unit in 2 heads
No. of spindles	6 spindles unit in 3 heads
	8 spindles unit in 4 heads
	10 spindles unit in 5 heads
Pirn length	150–250 mm
Bunch	0–15 m
Winding diameter	Max 30 mm
Spindle speed	Max 7000 rpm

Yarn speed	350 m per min		
Cone length	35, 45, and 50		
No. of traverse	6/10/13		
Electric motor 3 phase, 440 volts, 1440 rpm	0.5 HP for 2 spindles unit		
	1.0 HP for 4 spindles unit		
	1.0 HP for 6 spindles unit		
	1.5 HP for 8 spindles unit		
	1.5 HP for 10 spindles unit		
Machine size	Length	Width	Height
2 spindles unit	73 cm	108 cm	158 cm
4 spindles unit	143 cm	108 cm	158 cm
6 spindles unit	210 cm	108 cm	158 cm
8 spindles unit	273 cm	108 cm	158 cm
10 spindles unit	338 cm	108 cm	158 cm

2.12   References

- Conrad, R., and Siddall, G. J. (1953). An Experimental Study of Pirn Winding, *Journal of the Textile Institute Proceedings,* **44**(6), pp. 215–219.

- SenGupta, R. (1970). Yarn Preparation II, Mahajan Pub. Ahmedabad, p. 9.

- Lord, P. R., and Mohamed, M. H. (2005). Weaving: "Conversion of Yarn to Fabric", Woodhead Publishing Ltd.

- Sarabhai, V. A. (1978). Process Control in Weaving, ATIRA, Ahmedabad, p.15.

- Grallinger, C., "The Joy of Hand Weaving", D Van Nostrand Inc., New Jersey.

- Rude, S. (2005). Weaving: The Complete Process from Thread to Cloth, available at http://www.3springshandworks.com/Documents/Process.pdf

- Adanur, S. (2001). Handbook of Weaving, Sulzer Textile Switzerland, CRC Press.

3.1 Introduction

After finishing the yarn winding, selected cones are transported to next process, i.e. warping. Warping is an essential weaving preparatory operation in weaving unit. This is a very common say about warping, "good warping finishes fifty percent weaving."

3.2 Objective

The major objective of warping is to prepare a warp sheet of desired length containing a desired number of yarns that are wrapped on a flanged barrel in such a manner that tension in each yarn and density of yarn mass in the cylindrical assembly are maintained within a given tolerance level throughout the wrapping of the warp beam.

The process of warping is dedicated to the conversion of cones into a beam of given specifications.

3.3 Mathematical justification of warping strategy

In order to simplify the justification of warping strategies, let us assume that a woven fabric roll of 4 m width and 1000 m length is to be produced from warp yarn of 20 tex arranged on 25 yarns per cm in the fabric. This requirement can be fulfilled by the production of a beam in which 10,000 yarns are arranged parallel to each other in the form of a sheet. The length of each end forming this sheet should be higher than 1000 m to produce 1000 m finished fabric. During fabric manufacturing and during fabric processing, some major factors suppresses the warp yarn length. These factors are crimped path of warp yarn in fabric, contraction of fabric during subsequent relaxation processes, and wastage during fabric manufacturing. With the help of previous experience and literature available, an additional length of warp yarn has to be decided at this point. Approximately, 6% contraction in warp threads takes place due to crimped path which will add 60 m length in warp threads, and then 5% contraction in warp threads takes place due to fabric shrinkage during processing that adds 53 m length. Warp thread wastage remains around 0.63% which counts 7 m length. In this way an additional 120 m extra length is required; therefore, the final length of the sheet of 10,000 yarns must be kept to 1120 m. The mass of such warp sheet would be 224 kg only. About 2 kg cone of 20 tex yarn would contain 100,000 m of yarn. Therefore a minimum of 112 such cones are required to produce the target warp sheet. However, 112 cones will provide only 112 ends in place of required 10,000 ends.

In order to generate 10,000 ends, warp sheet of 1120m has to be withdrawn from each cone.

In order to generate 10,000 ends from 112 cones, one has to take out 1120m from each cone in the form of a sheet and then break the yarns from the cones and take out another 1120 m in the form of a sheet and repeat this process 89.28 times. The sheets generated at each moment, when superimposed or placed side by side, would produce the sheet of required dimensions and specifications. However, the problem is clear that 178.57 is not a whole number and therefore this solution is not practically feasible. The number 10,000 has to be therefore split up into combinations of cones (A), and beams (B) which will provide a realistic requirement of number of cones (A) and number of times the yarn has to be broken from the cone to restart the warping process.

Table 3.1 The possible combinations of cones and beams.

S. no.	Combination A × B	Details of combinations
1	10000 × 1	Each cone would contain 1200 m yarn length, weight 24 g
2	5000 × 2	Each cone would contain 2400 m yarn length, weight 48 g
3	2000 × 5	Each cone would contain 6000 m yarn length, weight 120 g
4	1000 × 10	Each cone would contain 12,000 m yarn length, weight 240 g
5	500 × 20	Each cone would contain 24,000 m yarn length, weight 480 g
6	250 × 40	Each cone would contain 40,000 m yarn length, weight 960 g
7	100 × 100	Each cone would contain 1,20,000 m yarn length, weight 2400 g
8	50 × 200	Each cone would contain 2,40,000 m yarn length, weight 4800 g

First combination is impractical because production of 24 g cone is industrially incorrect. To form the beam of 10,000 ends, creel of 10,000 capacity is required which is near to impossible in industrial conditions.

Second combination is also impracticable because 5000 cones of 48 g each are required.

Third combination is also impracticable because 2000 cones of 120 g each are required.

Fourth combination is also impracticable because 1000 cones of 240 g each are required.

The combination of 5th and 7th can be considered because both have few advantages. If the solution 5 is accepted, then the yarn has to be broken only 20 times and a creel of capacity 500 is required which is normally available in industry. The6th combination suppresses the creel capacity, but thread breakage number increases (40 times or 40 beams will form first which will combine by beaming process). The 7th combination is also acceptable, which requires the use of a reasonably small number of cones of standard dimensions. Cones of 2.4 kg weight are very popular in textile industry these days (Banerjee 2008).

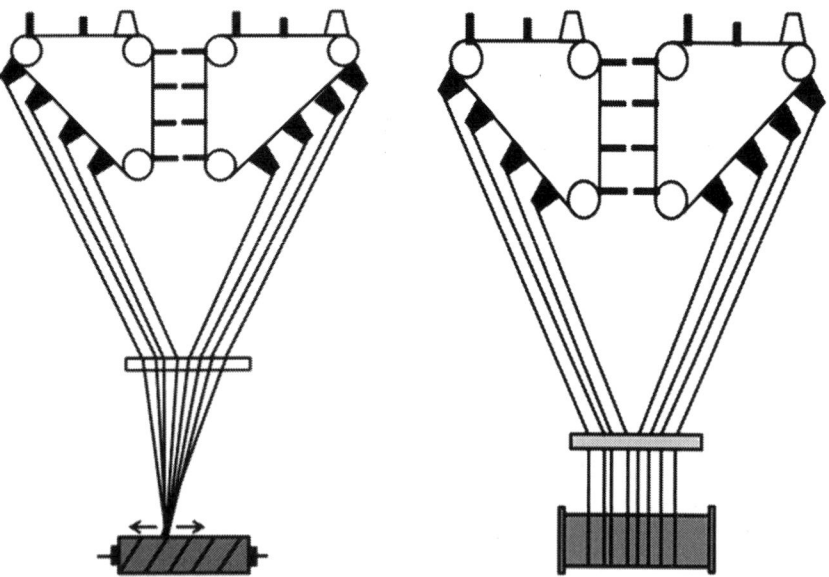

Fig. 3.1 Ball warping   **Fig. 3.2** Beam warping

The two solutions (5th and 7th) necessitate entirely different methods of warping, namely

- *Beam warping* – This is for solution 5 in which a creel of capacity 500 is required. 20 warpers beam will form first which will combine together by beaming process to get warp sheet of 10,000 ends.
- *Sectional warping* – This is for solution 7 in which a creel of 100 capacity is required, which will save the space requirement of creel section also. In this case, 100 sections of 100 warp threads each will form to get the sheet of 10,000 parallel ends. Finally, all 100 sections will transfer a beam to get warp sheet containing 10,000 ends.

The requirement of various colored threads in the fabric may also guide the choice between the two warping systems. Complex pattern of colored strips are easily obtained by sectional warping processes. In case of double-folded yarn for terry pile fabric manufacturing, warpers beam can be produced directly by sectional warping when sizing of warp beam is not essential.

3.4   Elements of the warping systems

The elements of a beam and sectional warping machine are shown in Figs. 3.3 and 3.4, respectively.

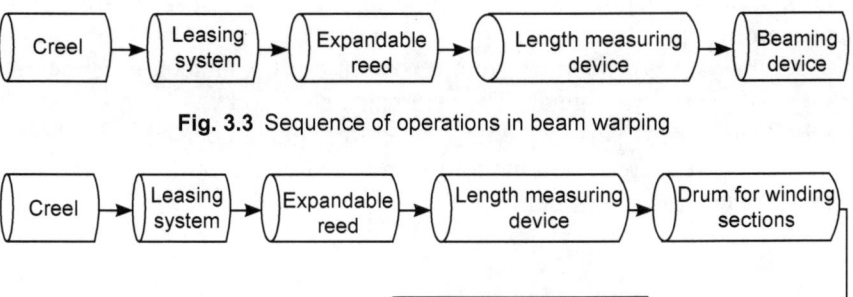

Fig. 3.3 Sequence of operations in beam warping

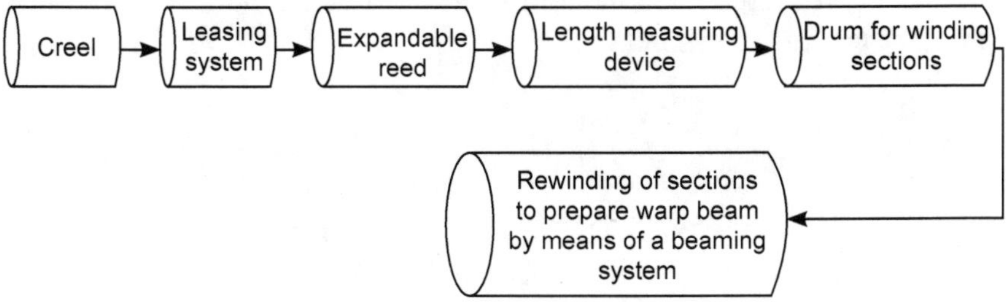

Fig. 3.4 Sequence of operations in sectional warping

Both beaming systems will produce warp beams; however, the beam from sectional warping system is complete in all respects whereas the number of warp yarns in the beam from direct beam warping system is a fraction of that targeted number of yarns in the final beam. Hence, a number of such beams have to be assembled in the subsequent sizing process for producing the final beam. Referring back to the example quoted at the beginning under combination no.5, 20beams have to be assembled in the sizing machine. In the sectional warping system adopted for the combination 8, 200 sections of 50 yarns each are wound side by side on the drum at first and then all the threads from these 200 sections are combined in a sheet form and rewound on to a beam. The beam warping system is also known as direct warping, as the warp beam is produced directly from cones arranged in the creel. Consequently, the sectional warping system is referred to as indirect warping as the warp beam is produced not from the cones in the creel but from the sections wrapped onto the sectional warping drum.

3.4.1   Various parts of warping machine

Warping machine is made of various essential parts like creel, measuring roller, open reed, head stock, and warping drum, etc. Various parts of warping machine are discussed below (Sarabhai 1978).

Creel
Creel is the most space-occupying part of the warping machine. There is a three-dimensional arrangement of pegs to hold the cones. The design of each peg is dedicated to hold the cones by gripping the inside wall of the shell of a cone securely, and be strong enough to support a cone in space at varying angles from neighboring cones.

As the yarn is withdrawn from a cone in creel, a balloon is formed in between cone and yarn guide. The yarn guide also works as balloon-control device. The yarn is passed further through tensioning device, and a thread detector (broken thread stop motion) is mounted on the frame as integrated part of the creel as shown in Fig. 3.5.

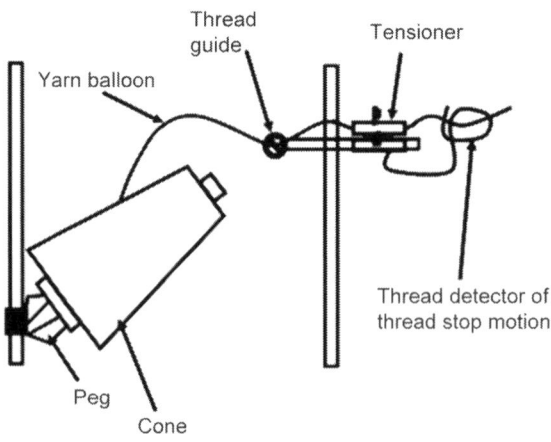

Fig. 3.5 Yarn tensioning device on warping creel

A creel that can accommodate 300 pegs must therefore have 300 sets of thread guides, tensioner, and thread detectors. Each peg occupies a different location in the creel. In consequence of that, the yarn path between the expandable reed and the peg is different for each yarn. This path difference results in varying tensions among the yarns, which vary with the yarn withdrawing angle and design of the creel.

V-shape creel

A creel in its plain view (Fig. 3.6) is usually V-shaped with the yarns emerging from the outside of the two arms of the "V" and converging at the expandable reed. Each arm of the "V" is in fact shaped like a rectangle. Thus two rectangular blocks are arranged in "V" form in such a creel, which contains a very large number of cones, and is therefore suited for beam warping machines. On the other hand, one rectangular block is sufficient to serve as a creel for the sectional warping machine. However, the creel of the sectional warping machine must be equipped with a motor-driving system for providing a traversing motion to the whole creel during winding of a section on the drum. Similarly, it should be possible to relocate the creel to a new position at the beginning of winding a new section on the drum. In fact the creel of a sectional warping machine has to be mounted on trucks that would permit the creel to move from one end of the winding drum to the other.

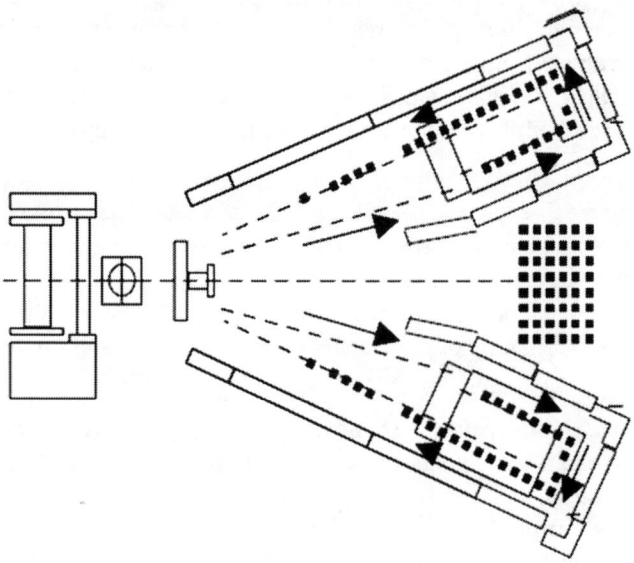

Fig. 3.6 V-shape creel

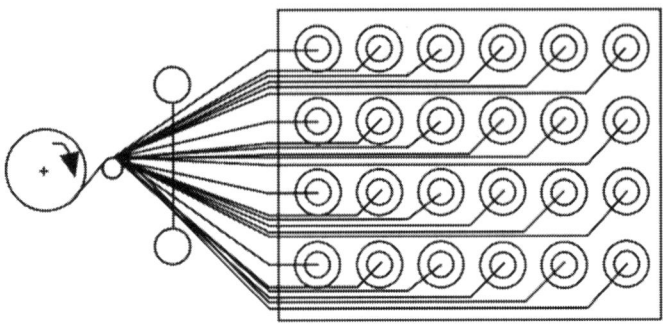

Fig. 3.7 Rectangular creel

Basically, four types of creels are available for the beam warping system. These are truck, swiveling, continuous chain, and magazine creels.

Rectangular creel
Rectangular creel is appropriate to serve as a creel for the sectional warping machine. In case of rectangular creel, all pegs are arranged in perfect rows which form a rectangular shape of complete block. Equal numbers of pegs are arranged in each row as shown in Fig. 3.7.

Truck creels
Truck creels employ reserve creels, which can be pushed in the place of the exhausted creel to save on the creeling time, i.e. the time required to replace empty cone shells from the pegs by new full cones. In case of very large size truck creel, the handling of trucks remains more difficult than swiveling creel because peg assemblies can be swiveled in and out in groups.

Swiveling creel
In swiveling creels each peg assembly carries two sets of pegs: one set carrying the cones from which the yarns are being withdrawn by the warping machine, and the other set carrying reserve cones. On exhaustion of cones, one has to swivel the pegs and bring the loaded pegs in working position, thus saving on time. During the running of the machine, the exhausted cones are replenished.

Continuous chain creels
In continuous chain creels, the frame of pegs carrying the exhausted cones are moved by a chain-drive away from the outer walls of the arms of the "V" to its inside position, while the reserve framework carrying full cones are driven simultaneously from the inside position to the outside.

In all the three examples cited, time is saved in replenishing the supply package. However yarn from the new package has to be tied (knotted or spliced) with the yarn unwound from the previous package, the tail end of which should be around the thread guide in the creel. This additional time is saved in magazine creels.

Magazine creels

In magazine creels, individually swiveling pegs are employed and the tail end of yarn emanating from the cone being unwound is tied in advance to the leading end of the reserve cone. Upon exhaustion of a cone, the pegs are swiveled and the warping process can be restarted immediately.

The swiveling, continuous chain, and magazine creels carry twice the number of cones required for warping. Hence space and accessibility become constraints if a large number of cones have to be used in warping. The truck creel needs on the other hand space for the reserve truck. Automation has taken place in creels in terms of threading-in of new yarns through tensioners and thread detectors. In the event of an end break, automated ladders move into the relevant position on the creel to assist the operator and lower the machine downtime. Modern tensioners employ feedback control systems to reduce the downtime.

Generally, two types of warping processes are used in textile industry. Both processes have their own advantages and disadvantages.

Table 3.2 Comparison between high speed warping and sectional warping.

Beam warping or direct warping	Sectional warping
Machine can run at very high speed	Machine can run at slow speed
Useful for manufacturing common fabrics in large quantity	Useful for manufacturing special fabrics and fabrics with special effects
Useful for producing warp beam using large amount of warp yarn	Useful for producing warp beam using small amount of warp yarn
Warp sheet of very high length produced	Warp sheet of short length produced
Weavers beam is produced after sizing	Weavers beam can be produced directly after beaming, if sizing is not essential
Generally, single yarns are used	Generally, plied and twisted yarns are used
One stage process	Two stage process
Warping cost remains low	Warping cost remains high
Easy to minimize yarn- to-yarn tension variation	Difficult to minimize yarn-to-yarn tension variation
Generally, cotton, wool, worsted, and linen yarns are processed	Mostly, silk and manmade fibers are processed
Creel capacity remains high	Creel capacity remains low
Cone or cheese packages are used	Flanged bobbins are used in case of silk

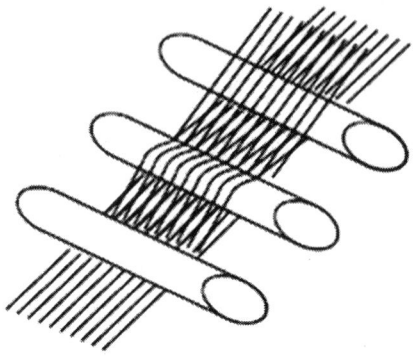

Fig. 3.8 Concept of leasing of warp yarn sheet

3.5   Leasing system

Leasing is a method of segregating warp yarns from neighboring ones and maintaining its location in the warp sheet as shown in Figs. 3.8 and 3.9. This assists the operator (the warper or the weaver) in easily locating the broken yarn and prevents crossed yarns mending in the sheet. Simple end-and-end leasing involves splitting the warp sheet into two layers of even and odd numbered ends. On a warping machine, this is carried out with the help of a leasing reed. A band or a rod inserted between these two layers would show either the odd-numbered or the even-numbered yarns passing above or under the band or rod. In order to create a neat crossover line, another rod or band is subsequently introduced between the two layers with the order reversed. In this way, neighboring yarns are forced into complete wave cycles of 180° phase difference. Identifying or locating the yarns at the crossing points of these waves is relatively simple. Leasing is also required for another purpose, namely separating yarns from each other after the warp sheet has been sized and dried. Leasing becomes helpful during drawing-in process also.

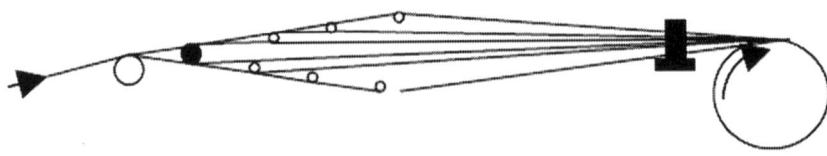

Fig. 3.9 Concept of leasing (side view)

3.5.1   Significance of leasing

A warp yarn sheet is completely encased in size film and cannot be used in the subsequent process until the yarns' individualization. When attempts are made to split the sized sheet into odd and even yarns, a huge amount of energy is required suddenly, and consequently, uneven distribution of stress and size across the warp sheet takes place. Anisotropic distribution of stress in warp sheet leads to rupture of yarns and peeling-off of size film. Hence multiple leasing is advised depending upon warp yarn density (closeness of warp sheet) to split the sheet into multiple layers, say 6, 8 or 10. In this way, every 6th, 8th or 10th yarn is separated from the main body of the warp sheet and a lease band is introduced in each gap. After the drying of sized warp sheet, careful and gentle tugging at such bands is carried out to segregate the individual layers in steps in order to individualize the yarns again. In few typical cases where sizing and multiple leasing are not required, an end-and-end leasing solves the purpose. The preferred time of leasing is at the beginning of the warping process, but the bands for the multiple leasing are introduced in the sheet either at the beginning (sectional warping, beam to be sized) or at the end (beam warping, beam to be sized). In any situation, the leading edge of the sheet of the warped beam would exhibit the multiple leases.

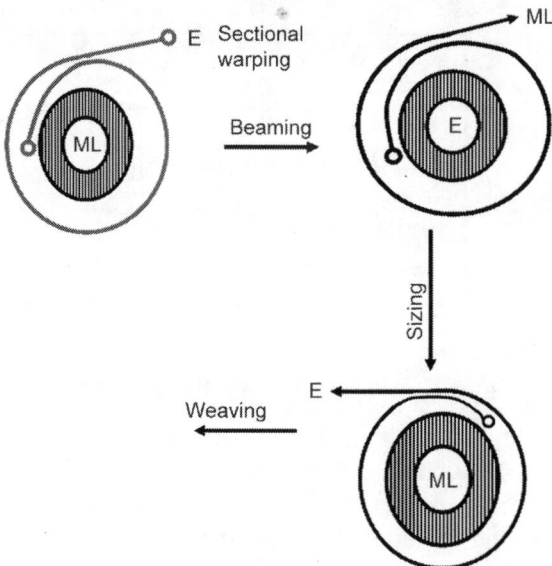

Fig. 3.10 Sequence of end-to-end multiple leasing

Automatic leasing systems illustrated in the Fig. 3.10 enable quick and fault-free leasing, especially for multiple leasing.

3.6   Expandable reed and length measuring system

The warp sheet from the creel converges into the expandable reed, through the dents of which the warp threads pass. As the reed wires can be moved either closer to or away from each other, the distance between the dents can be adjusted to the desired value of yarn spacing in the warp sheet. Thus, the density of the warp yarn sheet is set at the expandable reed. The length of the sheet in a warp beam is crucial from the point of wastage minimization. Therefore, the length measuring unit becomes a very important part of the warping machine. The measurement method and the accuracy of the same vary over a wide range.

3.7   Drum for winding sections on sectional warping machine

The drum on a sectional warping machine is a hollow cylinder with conical flange on one end, as shown in the Fig. 3.11.

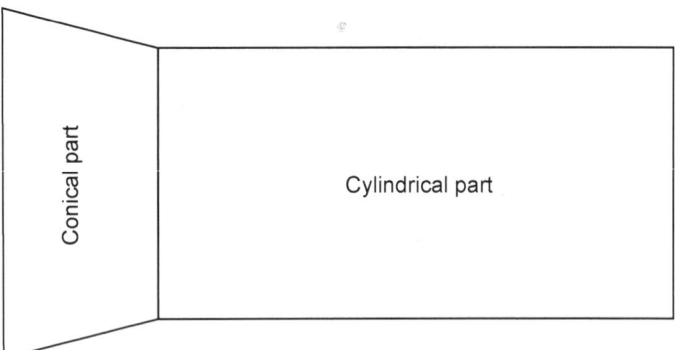

Fig. 3.11 Drum of sectional warping machine

The leading edge of a section of warp sheet is hooked on the drum surface at an appropriate location, and then as the drum rotates, the section starts getting wrapped on the drum surface. As the winding continues, the number of layers of warp on the drum would start increasing, raising steadily the effective instantaneous diameter of the drum at which a layer is being wound. The diameter of a bare drum being around 0.8 m (for circumference = 2.5 m) to 1 m (for circumference = 3.14 m), a substantial length of a warp section can be wound on the drum without causing a significant change in the effective drum diameter, and hence in the speed of winding, even if the drum rpm is kept constant. The speed of winding in a modern warper can be as high as 1000 m/min.

If the layers of a section were allowed to be wrapped on the same location of the drum, then after a while the sectional view of the warp sheet would appear like a rectangle and the two edges of the warp sheet would have a tendency to collapse in the absence of any lateral support. To prevent this from happening, the section is given a lateral shift or a traverse during the winding process, such that the cross section of the warp sheet wound on the drum appears like a parallelogram instead of a rectangle. The conical flange of the drum supports one edge of the parallelogram while the other edge appears tapered. This tapered edge needs no support and indeed provides support to the inclined edge of the next section as shown in Fig. 3.12. The conical flange of the drum plays therefore a vital role in sectional warping systems.

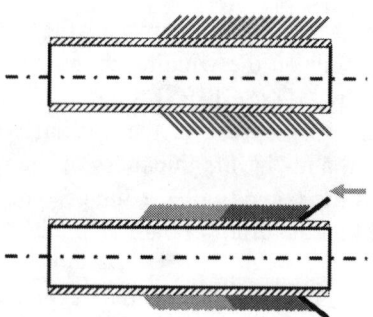

Fig. 3.12 Principle of conical flange

Winding drums are manufactured either with fixed conical flanges or with flanges that can be adjusted for variable conicity. The conicity angle of the fixed flanges is in the range of 7–14°, while the conicity of the other type can be varied over a much wider range. In the variable conicity drums, plates hinged at one end on the drum body are arranged along the circumference of the drum. By raising the free ends of the plates simultaneously away from the drum by equal amount, the conical surface is created. This surface however would not be as smooth as the surface of a fixed conical flange. Moreover, as one moves away from the hinge of a plate towards its tip, the gap between edges of the neighboring plates would keep on increasing. These two factors result in stress concentration on the yarn segments in contact with or very close to the flanged surface and also a difference in length between yarns wound on the flange and those on the main body of the drum at the corresponding diameters. This is also illustrated in the diagram. Thus, fixed conicity flanges would be preferred for delicate and sensitive yarns. However, the variable conicity system is more versatile to process a wide range of yarns as depicted from following facts.

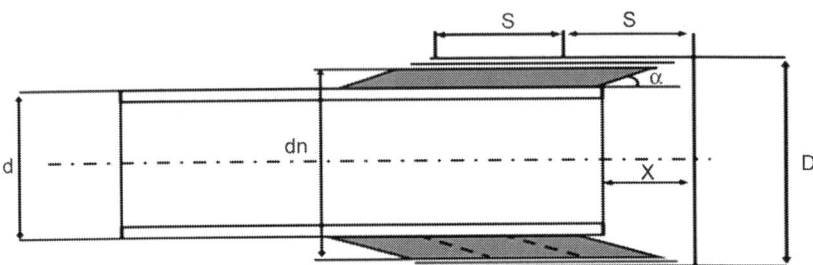

Fig. 3.13 Geometry of the sections and conical drum of sectional warping

In Fig. 3.13, each parallelogram represents the cross-sectional view of a section of warp sheet built on the winding drum. The area enclosed by any parallelogram is given by s. [$\{x^2 + b^2\}^{1/2}$]. sin α

In this expression, "s" represents the section width, "x" the traverse imparted to the section, and "b" the thickness of the resultant yarn sheet on drum, while "α" gives the conicity of the flange. The length of the sheet of thickness "b" wrapped on the drum of bare diameter "d" can be expressed as [π (b + d)].

Hence the volume of the yarn sheet can be expressed as

$$[\text{s.} [\{x^2 + b^2\}^{1/2}] . \sin \alpha] . [\pi (b + d)]$$
$$= [\text{s . x} \{1 + \tan^2 \alpha\} 1/2. \sin \alpha] . [\pi (b + d)]$$
$$= \pi. \text{s. x } (b + d) \tan \alpha$$

The width of a section is given by the warping requirements (number of ends per inch, and total number of ends in a section), which cannot be treated as a variable while adjusting the settings of winding machine. Moreover for obvious reasons, it is desirable to maintain the value of "b" as small as possible. Thus with fixed conicity flanges, only one variable traverse will remain available. The traverse can be adjusted for adjusting the volume to be wound, while with variable conicity drums both traverse and the cone angle can be adjusted. Besides the required volume of the warp sheet on the drum, the other factor affecting the traverse would be the yarn diameter. Clearly, thicker yarns would require a section to be traversed by a higher amount (yarn traverse at high traverse rate). Thus with fixed "b" and "α", the volume constraint may yield a value of "x" which may not be in agreement with the thickness requirement of the yarn. Hence if a wide range of yarn thickness is to be processed, then the variable conicity flange is advantageous. The value of "b" would be governed by the thickness of yarn, the yarn spacing, and also the tension applied to the section during winding. Higher yarn thickness, lower yarn spacing, and lower yarn tension would result in a higher value of "b" for the same length wound. Thus, a database would be needed to relate warp length of given specification of warp and winding tension with "b". Nevertheless, the following relation $b = x \tan \alpha$ implies that in the event of a fixed value of α, the traverse has to be accurately adjusted to get the exact value of "b". In the event of flexibility in choosing "α", the accuracy of setting would naturally be better. The winding tension, as discussed above, affects the value of "b" and therefore, the values of "x" and "α". In sectional warping, the importance of winding tension has another dimension. Normally the first section is wound from cones having the highest diameter and with every section the diameter of cones keeps on diminishing. Because of the ballooning effect, the unwinding tension keeps on increasing as the cone diameter diminishes. Thus the sections on the drum would be wound under progressively higher tension, resulting in a progressively diminishing "b" from the first section to the last. As the value of "x" is same for all sections, different taper angles of the supporting side of each section would result in the creation of instability in the warp sheet on the drum. Moreover during the next process of beaming-off, where all the yarns of the final warp sheet are wound on to a beam, large tension difference across the sheet would materialize. Such a beam would be difficult to size or weave. Hence, tension-compensating systems using feedback control loops are installed on modern sectional warping machines. Such machines employ servomotor-controlled yarn tensioners.

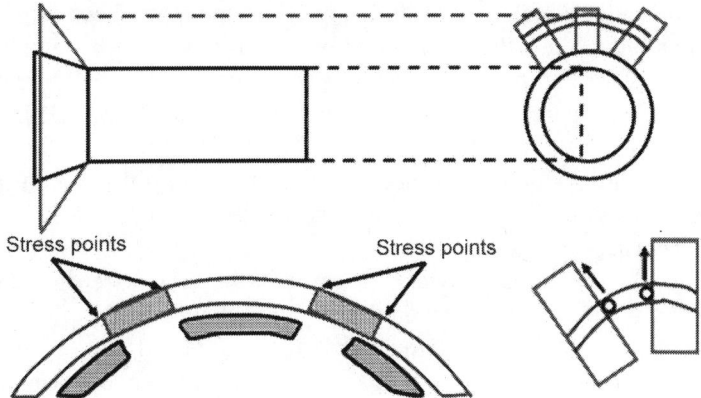

Fig. 3.14 Concept of variable conicity drum

Thus during winding the sections on the drum, the length of yarn and also the tension are monitored closely so that these two variables remain within tolerance limits for each section. Drum with fixed conicity is preferred for winding delicate yarns, which deform easily, i.e. yarns of low modulus and high extensibility. Figure 3.14 illustrates this point.

The conical flange of a variable conicity drum is made up of fin-like bars, each of which is hinged along the periphery of the corresponding drum end. Rotating the bars about their hinges would vary the flange conicity. Two such positions are shown in Fig. 3.14. As the bars are swung up for increasing the conicity, the distance between the adjacent edges of any two neighboring bars would grow as one move from the base to the tip of a bar. The segments of yarn lying on these bars would follow a curved path, while those lying in the gap between adjacent bars would be straight. Hence the net circumferential length along the bars would be always less than the one corresponding to a circle of equivalent diameter. Accordingly within each yarn layer wound above the base of the bars, a length difference would be observed among yarns wound on the flange and those wound on the solid drum surface. Moreover, this length difference would grow as one move to higher diameter. This length difference between the yarns of the same sheet would result in the permanent deformation of the shorter yarns during the subsequent beaming process. Over and above this deformation, the curved segments of a yarn are supported along their line of contact with the bar surface while the intervening straight segments are not supported at all, causing pressure points to develop at the edges of the bars where the curvature changes abruptly from a finite value to zero. Larger the gap between adjacent bars, more pronounced would be these zones of stress concentration. Quality of sections wound on the drum is also affected by the direction of rotation of the drum. The drum in the adjacent figure rotates in anticlockwise direction while the yarn sheet coming from the left is laid to the winding-on point on the drum surface by a guide roller. The guide roller and the winding-on point are quite close, and therefore the intervening segment of yarn sheet is constrained to follow a definite path. The top of the sections being wound is also fully visible and accessible for any intervention, which would not have been the case if the drum was being rotated clockwise. Furthermore, the automatic insertion of size-splitting element during the multiple leasing process is facilitated by the location of the guide roll. If the drum were to rotate in the clockwise direction, then the corresponding guide roll would have needed shifting during the leasing process, causing a disturbance to the threads.

A critical aspect of warping relates to the interdependence amongst the distance between the creel and drum, and the speed of winding and the braking torque that can be applied on the drum. When yarn breaks during warping, the drum has to be stopped within less than one revolution as otherwise the broken end would get embedded in the subsequent layer. In that situation, drum has to be revolved in reverse direction. During drum reversing, yarn entanglement takes place; hence, this system is not feasible in absence of any thread-accumulating device. Reversing the drum for finding out the broken end in the absence of any accumulating system would create entanglements in the yarns and hence is not feasible. Thus for larger winding drum with higher speed, more efficient braking system is required. A higher distance between the creel and the winding drum would ease the situation; although, it also means a higher space requirement. Having wound the first section to the desired length, the yarns are cut after introducing the lease band, and the end of the section securely tucked-in within the warp sheet. The other end of the yarn is securely gripped in front of the reed assembly. Subsequently the reed assembly and the creel are shifted to the starting position for the next section, lease band inserted into the tip of the new section, and winding on the drum repeated. An automatic leasing system becomes helpful to reduce downtime.

3.8   Sectional warping

Sectional warping process is used to produce stripped effect on fabric. Small size creels are sufficient to fulfill the requirement of sectional warping process. Generally, V-shaped or rectangular creels are integrated with sectional warping process. In this process, a section of small number of packages (10–300 or more) are mounted on creel and all threads are threaded through tread guides and tensioners before reaching on guide rollers, then expandable creel head, and finally at beam. All threads are entrapped in holes made on beam surface at appropriate position and then the beam starts to revolve. Machine is set to stop after a fixed number of beam revolutions. It depends on the required length of warp sheet on weavers beam. After completing the exact number of revolutions machine stops, then expandable reed head is shifted to its right and shifting depends on warp yarn density or warp yarn spacing in cloth. All threads are cut and then entrapped similarly in next slot on beam surface and winding of next section starts. Total number of sections depends on total number of threads required in cloth, and total number of threads in one section. After completing total number of sections, the beaming process starts. In beaming process, the last end of all sections is simultaneously reached on double flanged warpers beam and then starts the beaming process. During beaming process, power is given only to drive the double flanged warpers beam, and sectional beam is driven only by pulling force of all threads of all sections which are unwinding from its surface. During this process, tension uniformity is a major issue. Skilled warpers are needed to do this work precisely.

Table 3.3 Comparison between beam warping and sectional warping process.

Beam warping or high speed warping or direct warping	Sectional warping
Opt to produce common fabric in large quantity	Opt to produce stripped fabric or where creel capacity is limited
Opt to produce warp beam from large amount of threads	Opt to produce warp beam from small quantity of threads
Production speed is high	Production speed is low
Requirement of creel capacity is high	Requirement of creel capacity is low
Generally beam warping followed by sizing	Sectional warping can be followed directly by weaving
Generally, single yarn is processed	Generally, ply or twisted yarn is processed
Higher length of warp sheet is produced	Lower length of warp sheet is produced
Generally, cone or cheese bobbins are used	Generally, double flanged bobbins are used
One stage is sufficient to produce weavers beam	Two stages are involved to produce weavers beam
It is easy to maintain uniform yarn tension	It is difficult to maintain uniform yarn tension
Process is used popularly for cotton , woolen, worsted, and linen yarns man made yarns	Process is used popularly for silk, and
Less expensive process due to high speed, and one stage process	Expensive process due to low speed, and two stage process

3.9   Beaming system (thread transfer from sectional warping drum to warpers beam)

This is an indirect beam formation system. In this indirect system of beaming, the ends of sections wound on the drum are collected in a sheet form and then wrapped onto a double-flanged barrel to form the warp beam. The beam is rotated by a spindle drive, and the tension created in the warp sheet due to rotation rotates the winding drum, unwinding all the sections simultaneously. The torque required to rotate the winding drum is generated at the moment of inertia of the drum and also the other resistances such as the braking force applied on the drum and the friction in the bearings, etc.

Therefore, the beaming speed remains low, i.e. around 300 m/min. Judicious density of the warp beam is maintained with the help of press-rolls and the angular velocity of the beam which is gradually reduced with build-up of the diameter. Warp beams of diameter more than 1 meter are produced on the beaming system.

As the layers in each section are wound with a traverse, a matching traverse in the opposite direction has to be imparted to the winding drum, so that the edges of the warp sheet coming out of the drum always remain parallel to the flanges of the warp beam.

In case of a direct beam warping machine, small number of yarns, like 250–500, is pulled out of the creel and wound on a flanged barrel directly. Both surface and spindle drives are applicable in direct beam warping. A very small resistance of unwinding is exhibited in this process; the beaming speed may reach up to 1000 m/min. The beam can be surface driven or spindle driven. The surface-driven system results in a constant winding speed, while the spindle drive would necessitate additional control system for reducing the angular velocity of the beam with rise in beam diameter. Press rolls are employed on the beam for generating the required hardness. In the event of a thread break in a surface-driven machine, brake must be applied individually on the beam, the driving drum, and the press rolls, and also on the measuring roll. This can be easily achieved with a thyristor-controlled driving system, which prevails in a spindle drive. Therefore, the modern direct warping machines have employed spindle drive. Surface drive is also harsh for the warp, especially during starting and stopping.

3.10   Developments in warping by Tsudokama

3.10.1   KB 30 and KB 30-SH beamer by Tsudakoma

The KB30 and KB30-SH beamers meet market demand with its high adaptability when fabric changes, responding to the needs for larger warp beam, wider fabric weaving or higher take-up tension winding. The KB30 and KB30-SH are assisted by a line-up of excellent peripheral devices also. Tsudakoma has marked a top-ranked delivery record of beamers all over the world, not only for sized yarns but also for non-sized warps. KB30 and KB30-SH beamers are modified form of the KB20 to achieve widening of the take-up tension range, lowering noise level at the head stock, and elevating easy operation with a compact design.

3.10.2   Incomparably stable take-up tension control

Tsudakoma developed a stable take-up tension control system to wrap big warpers beam at constant take-up tension. The AC vector motor directly drives a beam to take-up yarns while adjusting the tension automatically. The original tension feedback control based on the torque of the AC vector motor connected with the taking up roll stabilizes take-up tension. This also allows section of taper tensions in accordance with yarn characteristics.

- The tension control range is set between 300 N and 5900 N for taking up on a beam with flanges of 800 mm diameter (optional: ~7800 N, for 800 mm flange diameter) with superior stability in low tension take-up.
- Gears contained in an oil bath for driving beams to assure less noise generation.

3.10.3   Easy operation

The cradles on both sides to hold both ends of the beam are moved by independent motors to simplify the operations to mount or unload beams. When replacing a warpers beam, it can be automatically loaded or removed by an optional lifting device with a hydraulic system, in addition to the movement of the cradles on both sides.

3.10.4   Integrated control by MDS machine data station (optional)

By providing tension feedback control to an individual beam stand, the tension control for the beam stand is fully automated and the take-up tension control is totally integrated.

- By feeding operational conditions corresponding to the articles (processed yarn) into the MDS in advance, it becomes possible to set proper machine conditions to produce a beam with minimum fault and superior end-to-end tension homogeneity.
- Automatic tension control is also attainable, in case different tension is required for each beam stand (because of the difference in kind, thickness or quantity of yarns on section beam).

3.10.5   Stable and promising mechanism

Beam stand with powder brakes
- The tension is automatically controlled in relation to the change of section beam diameter by means of air-cooled powder brakes featuring uniform and stable torque characteristics, resulting in consistently uniform yarn sheet arrangement to ensure clear warp shedding on looms.
- A run-out preventer for machine stop and a powder supply for power failures is provided. It stabilizes the operation, keeping constant tension

not only during low-speed operation, but also at start-up, stop, and power failure. This run out preventer is applicable for warpers beams of flange diameters as large as 1000 mm.

• The beam stands are arranged so that each yarn sheet is taken up with even intervals to avoid snarling.

Optional

• The beam stand tension indicator on the front of the control panel, which enables centralized tension indication and control of yarns drawn from each stand, assures further stable operation of warp beam.

• For taking-up on a wide warpers beam, the beam stand with guide roll is available to avoid forcibly pulling yarns out of each section beam.

• Section beams of different widths can be placed on the adjustable width-beam stand on which both sides of the stand move separately.

• Weave pattern changes often invite finished warp beams to remain unused. Warp rewinding from such remained warp beams is also possible on this beam stand under stable and high winding tension assisted by powder brakes.

Tsudakoma made all these efforts to produce fault-free weavers beam for further processes.

3.11   Perfect beam

A warp beam can be said as perfect warp beam after incorporating all the following conditions.

• All warp ends must be wound straight and parallel to one another from beginning of beam to the end.

• Each end should hold its relative position in the warp with no "rolled", "crossed", "stuck", or "lost ends".

• Tension uniformity must be maintained on all ends.

• Density of the warp should be kept uniform from beginning of beam until finished beam.

• Selvage ends should be flat with the rest of the warp.

3.12   Sectional warping machine manufacturer in India

Prashant Gamatex Pvt. Ltd., Ahmedabad, is prime name among the Indian warping machine manufacturers. Lasertronic 1080/1100 is the leading model of Prashant Gamatex in recent years.

3.12.1   Features of Lasertronic 1080/1100

This machine consists of solid steel drum of 3.14 m circumference; automatic section-wise tension control system which has the following features:

- Separate warping and beaming system
- Solid steel drum manufactured with international standards with 3.14 m circumference
- Twin PLC system with closed-loop yarn tensioning monitoring used to control the tension in every section
- All major mechanisms controlled with dedicated servo motors with ball screws
- Laser-based contactless pitch measuring system
- An adjustable continuous beam oscillation system
- Drum is able to make reverse rotation if required
- Automatic leasing system
- User friendly hand-operating panel
- Beaming at constant tension throughout the process
- Large touch screen for easy data processing in colored mode
- Lost-end memory system for efficient tying-up
- Most of the motions controlled by servo motors
- Linear guide available under carriage system
- Interface printer device to get data sheet in printed form
- The available working width of1800–4200 mm
- Able to achieve warping speed of 0–800 m/min
- Able to achieve beaming speed of 0–100 m/min
- An efficient tensioning system integrated to achieve tension in the range of 0–450 kg
- Maximum 800 mm beam diameter can be achieved with 1000 mm flange
- Three roller warping table attached to maintain constant distance between rollers and warping point from first to last section
- A feeler roller available to control build up of cylindrical warp over full width

3.13   Latest developments in warping by KARL MAYER

KARL MAYER developed a warping machine for sample and running warping production named as Multi-Matic. Multi-Matic has the capability to produce up to 128 warp threads with a warping speed of 800 m/min. The development of Multi-Matic was dedicated to cover the gap between the

GOM 24 and Nov-O-Matic. The potential of Multi-Matic has been enhanced five times than GOM 24 in terms of number of yarn process ability. In case of Multi-Matic, the yarns are directly taken off from supply package on typical creel. The yarns are selected by means of yarn guides mounted in line of passage of yarn. The yarn selection system is based on computer intelligence that works in a programmed fashion for single thread selection. Yarn can be placed up to accuracy of 0.05 mm and with the sufficient acceleration of the motors. These features make Multi-Matic a useful machine to process a wide variety of materials from silk to polyamide multifilament yarns.

3.14   Latest developments in warping by Benninger

Benninger, a very well-known Swiss warping machine manufacturer launched a latest model "Versomat" of sectional warping machine dedicated to warping of samples and short warp lengths capable to produce a high quality warp beam with guaranteed efficiency. The Versomat allows warps to be produced economically using the smallest quantities of yarn, at the same time as offering the ability to manufacture work in normal production lengths. Versomat is equipped with following features to make it competitive in international market.

Latest technology
- This machine has an integrated reed headstock into the machine superstructure that allows very high production speed.
- Production of warp sections with a minimum width of 4 mm or minimum 12–14 threads depending on yarn fineness with maximum section width of 150 mm or 480–560 ends can be possible on this machine.
- The perfect tension between creel and drum can be ensured with the use of controlled section tensioner. Finally, a controlled beaming tensioning system achieves the requirement of high quality warp beam.
- The time for each separation at leasing point is 7 seconds.
- This machine is available with a working width of 2200 mm for traditional and 3600 mm for decorative and furnishing covering manufacturing.
- This machine is capable to maintain the warping speed of 750 m min−1.
- The leasing device is a part of machine and located in the machine superstructure. This technology reduces the downtime at the time of lot change.
- The machine is equipped with a laser-aided feed system along with section tension control that results in a correct package build-up.
- Optical monitoring of the thread separation at the warping reed during automatic insertion of the split cords (which are simply placed in the

warping drum as a part of the setting-up process) is extended in scope by the optical positional check of the separation slider. In consequence of this, an even cylindrical package build-up and the correct placement of the split cords, both of which are essential for perfect warp quality.

- This machine is suitable for the warping of silk yarns, multi- and monofilament yarns, and staple-fiber yarn from the finest cotton yarn to the coarsest woolen yarns.

3.14.1   Benninger has developed another sectional warping machine ERGOTEC

ERGOTEC is one of the latest sectional warping machines worldwide till date. Following facts make ERGOTECH a most competitive sectional warping machine.

ERGOTEC is equipped with a reed headstock with precise and automatic positioning of warp threads on warping drum. During section build-up, the evener roller ensures that the yarn count variations and yarn volume changes do not cause any density variations; thus guaranteeing a uniform cylindrical build-up. This is able to take reserve rotation. Correct section alignment is achieved by exact traverse of the reed headstock with correct section width setting. The ERGOTEC provides enormous advantages by controlling the section width because the effective section width can be measured at the measuring roller. Section marking caused by incorrect measurement of the section width can thus be effectively eliminated. Reed headstock feed and yarn tension has a decisive influence on warp quality. If during warping the reed headstock feed is changed, trapped ends at the section edge are unavoidable. At the same time, the unavoidable yarn tension variations lead to changes in section build-up, winding density, and section width.

This leads to the dreaded intermittent stripyness at the section repeat, which is especially noticeable in pastel-shade piece-dyed fabrics. Reed headstock feed and yarn tension sustain absolutely constant parameters with the ERGOTEC, and they do not change during the whole warping process. In this way it is absolutely certain that the same conditions exist from the first to the last section, regardless of warping speed or yarn package diameter.

The open yarns run from the creel to the measuring roller on the reed headstock. The rolling of ends in individual dents of the guide reed can be effectively stopped. Particular care was taken in the development of the ERGOTEC to optimize distance between the warping and beaming sides of the machine to assure quick reach of operator after appropriate operation research. When working without a waxing device, the minimum free warp-sheet run can be achieved without having a use of fluted guide roller. The braking system has been modified by using the warping motor in generator mode to control the warp tension more precisely. With this system, extraordinarily constant and extremely low beaming tensions are possible.

Thus low end effect yarn warps can be beamed with a defined beaming tension without problems. With this new system, the beaming tension control equipment is maintenance free, and during operation there are no squealing brakes or noise coming from cooling fans.

The energy generated by the warping motor is used to power the beaming motor. The installation has therefore a considerably reduced power requirement.

The steplessly adjustable traversing device and the digressively adjustable beaming tension is a standard feature of the ERGOTEC, and is the automatic stop for pre-stored lost ends, which also signals to the operator, that which section the lost end is in. It is thus guaranteed that no warp beam goes into the weaving department with any lost ends. It goes without saying that with the ERGOTEC all necessary additional devices such as static eliminator bars, waxing device, and beam pressing device are available to ensure optimal warp quality.

Productivity of the section-warping installation has been significantly improved. Thus the maximum warping speed of 1000 m/min can represent an increase of around 25% with improved beaming power and performance. These are however the smaller parts of the productivity increase. Therefore, much emphasis was placed on optimizing the individual work elements of the ERGOTEC.

The semi-automatic leasing device ERGOSPLIT is attached with ERGOTEC. The ERGOSPLIT was developed after optimizing the handling distances at the reed headstock and the drum. Hence, insertion of split cords (without the effort of opening and closing a guide roller) as well as the extremely fast positioning of the reed headstock and the drum requires very short warping times. The beaming process has also been further optimized. The simplest transfer of the warp sheet direct from the warping to the beaming side becomes possible. The transfer is quite rapid now with precise positioning of the drum to align the correct beaming position, as well as the extremely rapid opening and closing of the safety guard. The graphic touch screen terminal on the beaming side further simplifies the operation. All the relevant data for making a section-warped beam are stored and controlled by the total quality control (TQC) system.

Therefore, erroneous operation of the sectional-warping process has been excluded to make

ERGOTEC more user friendly.

In this connection, there are two points especially worthy of note.

- *Process plausibility verification (PPA):* ERGOTEC is equipped with PPA system which immediately and viably prevents entry of uncorrelated data. This gives the operator the security of always being able to enter data correctly, and with inexperienced operators, this renders time-consuming checking superfluous. Also during the warping and beaming operations, the operator's correct work sequence can be continuously monitored.
- *Wizard-aided programming (WAP) data entry*: This machine is supported by a WAP data entry wizard. This system ensures about the feeding of each data.

Furthermore, the following elements are equipped to enhance the workability and accuracy of this machine:

- *Automated end counting (AEC):* This system is used for automatic recognition of the number of running ends compared with the required number of ends per section. This system avoids the time-consuming counting of ends in the reed and possible errors due to working with different numbers of ends in each section.
- *Package quality monitoring (PQM):* It is an automatic yarn package quality monitoring system which works on signals after the third stop, the position of a problem bobbin within a pre-programmable time, which can then be changed.
- *Automatic leasing sequence (ALS):* The semi-automatic leasing device ERGOSPLIT effectively eliminates errors due to incorrect opening of the lease shed on the basis of automatic leasing-sequence system. The leasing sequence is preprogrammed and automatically carried out. The operator has no possibility to insert a lease cord incorrectly.
- *Automatic drum positioning (ADP) system:* With the help of ADP system, the drum is rotated in the shortest time to the exact leasing position and section start/end position. The chances of faults generation from uncontrolled drum rotation at exact leasing position are negligible now.
- *Automatic feed determination (AFD) system:* After the integration of AFD system, automatic predetermination of the reed headstock feed ensures that an absolutely cylindrical build-up is assured on the drum. The set feed remains constant throughout the entire warping process.
- *Section tension control (STC) system:* With the help of this system, the entire warping process is carried out under measured and controlled tension. This system assures that the warping speed or bobbin

diameter will not influence the thread tension throughout the process. This is an important prerequisite for a cylindrical warp buildup on the drum.

- *Automatic section alignment (ASA) system:* This ASA system determines the section width automatically and assures exact section alignment.
- *Process data storage (PDS) system:* With the help of this PDS system, the entire data for a warp can be stored in the data base and recalled when required. Thus, reproducibility and uniformity of the quality of warps can be assured. The initial entry of new warp data is possible at the production planning stage. The data can be downloaded from an external data base to the sectional warping machine at any time.
- With the help of TELELINK system, it is possible to connect the ERGOTEC to the Benninger service centre. This allows rapid and effective appraisal of the status of the installation. It is also possible that this system can directly send the intervention advice to the operator on the spot of possible incorrect actions and directly correct them. The clear and all encompassing presentation of individual components, as well as sensors and actuators on the graphic display significantly simplify the operation of the section-warping installation.
- Service and spare parts facilities are of decisive significance to the operational reliability and availability of the installation.

The machines must operate 24 hours per day and 7 days per week. If the installation is out of production, it is emphatic that it goes back into operation as quickly as possible so as not to compromise the time-table for downstream processes. In this connection, Benninger stands as a strong and supportive partner. As a worldwide operating company, response for spare parts and/or a technician is guaranteed in the shortest possible time. Competent technical advice and support, not only in the warp preparation department but also in the total process sequence of a textile manufacturing operation, is becoming ever more important. In this area, Benninger is the only worldwide company in a position to offer a wide ranging competent consultancy from the spinning department to the finished fabric. Benninger textile technological consulting ensures that Benninger installations operate optimally not only from the point of view of productivity, but also from the qualitative view point of the textile end product. This service extends from technical training of the operators "on the job" through repeated audits and bench marking of the individual departments, to comprehensive plant evaluation.

3.15   TW1000F warper by Tsudakoma

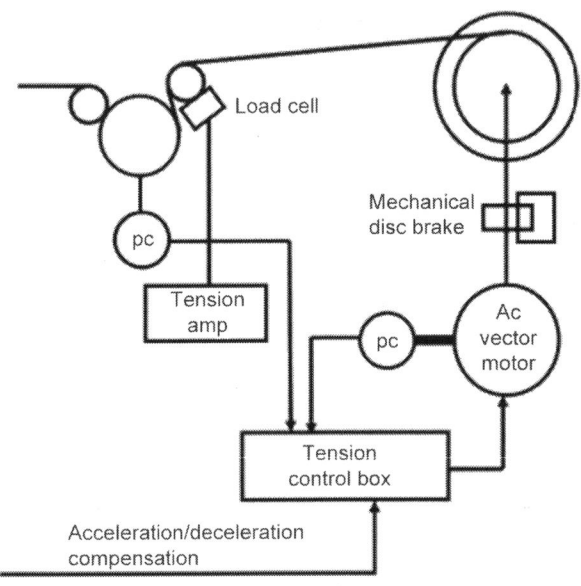

Fig. 3.15 Take-up tension system (TW1000F warper by Tsudakoma)

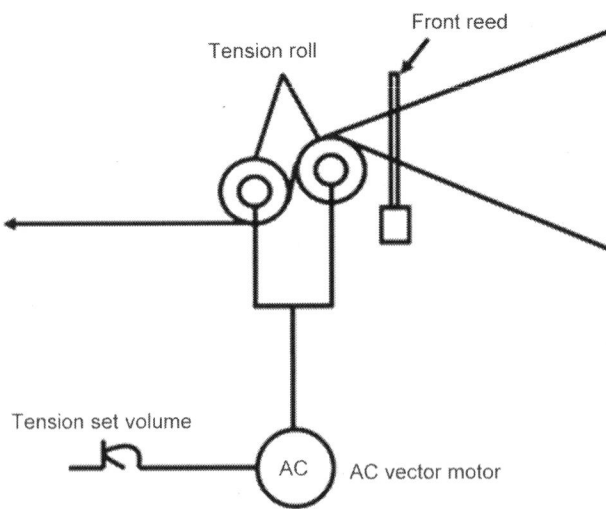

Fig. 3.16 Tension control system

3.16   Controls of end breaks in warping

Warp threads breakage in warping can be minimized by taking care of the four major points.

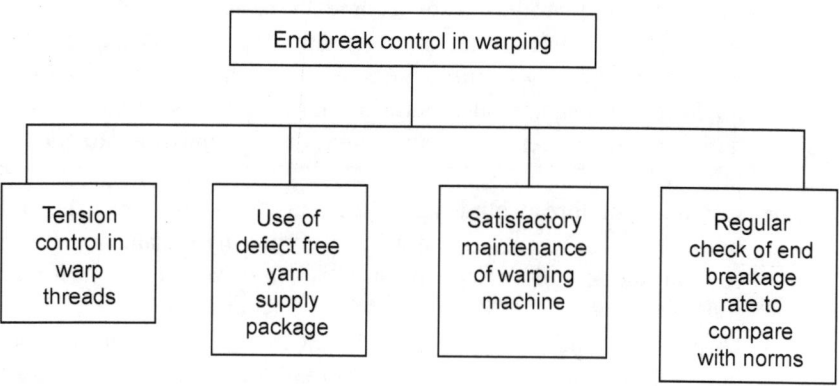

3.16.1   Tension control in the yarns

During warping process, two types of yarn-tension variation takes place (Rude, 2005).

- First type of tension variation takes place due to unwinding of yarn from different stages of the supply packages. This tension variation can be measured just after the first thread guide in creel section.
- Second type of tension variation takes place due to variation in yarn length from first thread guide in creel section and head stock, and angle of yarn from first guide and head stock.
- The tension variation of first type can be minimized by alteration in creel design in such a way that the distance between the package and the first thread guide is minimized, that will avoid the yarn balloon from rubbing the nose of the package.
- The tension variation of second type between the ends at head stock is minimized if the groups of neighboring ends on the beam are taken from the same column of packages in the creel.

3.16.2   Condition of warping machine

Working condition of warping machine also decides the end-breakage rate during warping process in case of both beam and sectional warping. The major points in machine conditions are given below (Sarabhai, 1978):

- Alignment of package holder with respect to first yarn guide in creel section should be properly checked and aligned with the help of special gauge. It is found that non-aligned packages register higher number of end breaks in comparison of aligned packages.
- Eccentric guide rollers also become the cause of end breakage. A number of guide rollers are used near the headstock in case of machine equipped with mechanical stop motions, and eccentricity of guide rollers becomes the cause of short-term tension variation of high amplitude. Hence, eccentric guide rollers must be attended at priority level.
- Condition of thread guides also influences the yarn-breakage rate because any deep cut in yarn guide significantly enhances the yarn tension due to sudden increase in the area of contact of yarn with rough surface.
- Dirty thread guides can be washed with carbon tetra chloride (CCl4), but CCl4 is a harmful chemical from environmental point of view because it affects the ozone layer.
- Warping machine should be cleaned carefully without blowing fly in the department that may wind with some other beam.
- While changing the yarn count and type, yarn tension must be checked by tension meter and adjusted, and it should be checked regularly.
- Stop motion and brake: High speed warping machine should stop after getting signal from any stop motion with $1\frac{1}{2}$ revolution of warping drum.
- Guide rollers should not be eccentric and eccentricity must be checked half-yearly.
- Creel fans should swing properly, otherwise the chance of fluff sticking enhances.
- All machine parts coming in yarn path should be checked weekly for any cut, because any cut will become the cause of yarn breakage.

3.16.3   Relative humidity and temperature

Optimum efficiency of warping machine can be obtained by keeping the relative humidity, 60% of the warping department. Lower humidity level becomes the cause of increase in yarn breakage rate, yarn hairiness, and fluff generation. When fluff passes with yarn and wind on warpers beam, it creates very serious problem during sizing. Very high humidity (higher than 70%) can again increase the yarn-breakage rate.

3.16.4   Use of defect-free supply package (cones)

The supply package for warping machine, generally a winding cone, should be free from defects. Supply package defects like patterning, hard package, soft package, cauliflower package, and jali on package become the cause of poor warping. These defects become the cause of end breakage and tension variation during warping process.

3.16.5   Regular check of end-breakage rate to compare with norms

End breakage during warping cannot be nullified, but there are few norms within the limits of norms in which end-breakage rate is permissible. Different mills have their identical norms to control the warping process, and few norms are optimized and published by different research organizations in India like Northern India Research Association (NITRA), Ghaziabad, Southern India Research Association (SITRA), Coimbatore, Bombay Textile Research Association (BTRA), and Ahmedabad Textile Industry's Research Association (ATIRA), Ahmedabad in India.

3.17   References

- Sarabhai, V. A. (1978). Process Control in Weaving, ATIRA, p.15.
- Lord, P. R., and Mohamed, M. H. (2005). Weaving: "Conversion of Yarn to Fabric", Woodhead Publishing Ltd.
- Adanur, S. (2001). Handbook of Weaving, Sulzer Textile, Switzerland, CRC Press.
- SenGupta, R. (1970). Yarn Preparation II, Mahajan Publications, p. 09.
- Rude, S. (2005). Weaving: The Complete Process from Thread to Cloth, available at http://www.3springshandworks.com/.
- Banerjee, P. K. (2008). Class notes, Professor Banerjee, P. K., IIT Delhi.

4

Yarn sizing

4.1   Introduction

Warp yarns are used to supply on looms where warp threads are subjected to repeatedly occurring various stresses like cyclic strain, flexing, and abrasion. These unavoidable stresses cause frequent warp breaks which consequent in the form of loom stoppages and finally, efficiency loss and fabric faults come in account. To minimize these problems, warp yarns are needed to be sized.

Warp sizing is an essential process to coat warp yarns with elastic film and bind fibers of yarns and lowering fluff or hairiness so that the yarns may become so strong that they can resist the mechanical strain in weaving process, and/or maintain or improve weaving efficiency. In other words, sizing is done to provide a protective coating and to lubricate the surface of the yarn to enable it to withstand the abrasive action which occurs in the weaving process.

These facts define that the sizing process is very important for a successful woven fabric manufacturing.

4.2   Objective

The main objective of sizing is to form a uniform coat over warp yarns and lay down the protruding fibers on yarn body that protrude out of its surface. After sizing, the strength and abrasion resistance of the yarn improves and hairiness decreases.

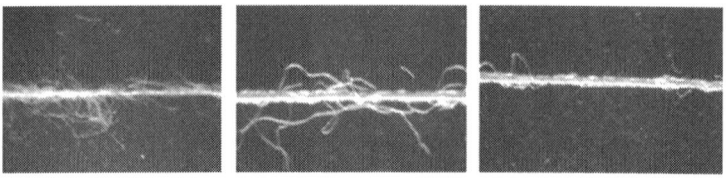

Fig. 4.1 Views of two-unsized and one-sized yarns

The sizing material should be selected in such a way that it could be easily removed during the process of de-sizing. The size material should penetrate the body of yarn to such an extent that the film gets a firm anchor. Excess penetration of size would not only mean more material consumption but also reduced yarn flexibility. A well-sized yarn has a higher work of rupture than the unsized one. The film of size surrounding a warp thread has to be elastic in tension and repeated flexing, while its surface should be well lubricated and should possess a high abrasion resistance.

It should not delaminate during the weaving process but should come away easily during the desizing process.

The film of size not only protects the neighboring warp threads on a weaving machine from rubbing together and getting entangled through formation of globules of fibers, but also elevates the work of rupture of the yarn, which is crucial for withstanding the weaving strain. Evidently, the recipe of size would depend on the nature of warp as also on the weaving conditions. A modern sizing machine can process yarns coming from about 25,000 ring spindles and feed 150 projectile looms of 3 m width. It is therefore evident that a typical weaving unit would possess a limited number of sizing machines, the maintenance and proper operation of which is very crucial to the entire weaving shed. From the foregoing it can be inferred that an understanding of the nature of sizing materials as also of the crucial process variable is central to ensuring a good sized warp.

4.2.1   Fluff binding

Fluff binding is a very important factor. Fluff binding prevents the yarns from clinging with each other. Fluff binding becomes very crucial in case of high-density fabric or in case of high-speed air jet weaving. Fluff-free sized yarns form smooth and clear shed during fabric manufacturing.

4.2.2   Size materials

The principal components of a sizing recipe are
- Adhesive
- Lubricant
- Antimicrobial age

4.2.3   Desirable bonds for sizing

In order to hold the protruding fibers, the number of bonding points of the sizing material with the fiber must be high. Type and number of bonds decide the cohesive strength of the sizing ingredient.

4.2.4   Cohesive strength

Cohesive strength is the total strength including elasticity of sized yarn to resist such mechanical strains posses in weaving process as tension, bending, rubbing, and shedding, etc. Cohesive strength is mainly derived by various bonds formed between size material and fibrous material.

More bonding has to be accomplished by weak bonding forces, which are discussed below.

Van der Waals' forces
These are very weak forces of attraction arising from the following:
- Momentary dipoles occurring due to uneven electron distributions in neighboring molecules as they approach one another.
- The weak residual attraction of the nuclei in one molecule for the electrons in a neighboring molecule. Stronger Van der Waals' forces are required to attract more electrons in atoms.

The more electrons present in the molecule, the stronger will be the Van der Waals' forces. Van der Waals' forces are single intermolecular forces operating between non-polar molecules; for example, Van der Waals' forces operate between hydrogen (H_2) molecules, chlorine (Cl_2) molecules, carbon dioxide (CO_2) molecules, dinitrogen tetroxide molecules (N_2O_4) and methane (CH_4) molecules.

Dipole–dipole interactions
The strength of dipole–dipole interactions is greater than Van der Waals' forces. They form interactions between molecules that have permanent polar molecules (net dipoles); for example, dipole–dipole interactions occur between SCl_2 molecules, PCl_3 molecules and CH_3Cl molecules. If the permanent net dipole within the polar molecules results from a covalent bond between a hydrogen atom and either fluorine, oxygen or nitrogen, the resulting intermolecular force is referred to as a hydrogen bond. The partial positive charge on one molecule is electro statically attracted to the partial negative charge on a neighboring molecule.

4.3   Hydrogen bonds

Hydrogen bonds generally result from electrostatic attraction between electronegative atoms (such as O or N) and a hydrogen atom that is bonded covalently to a second electronegative atom (such as O or N).

Hydrogen bonds occur between molecules that have a permanent net dipole resulting from hydrogen being covalently bonded to fluorine, oxygen or nitrogen. For example, hydrogen bonds operate between water (H_2O) molecules, ammonia (NH_3) molecules, hydrogen fluoride (HF) molecules, and hydrogen peroxide (H_2O_2) molecules, alcohols such as methanol (CH_3OH) molecules, between carboxylic acids such as acetic acid (CH_3COOH), and between organic amines such as methyl amine (CH_3NH_2). Hydrogen bonds are stronger inter-molecular forces than either Van der Waals' forces or dipole–dipole interactions since the hydrogen nucleus is extremely small and positively charged, and fluorine, oxygen and nitrogen being very electronegative so that the electron on the hydrogen atom is strongly attracted to the fluorine, oxygen or nitrogen atom, leaving a highly localized positive charge on the hydrogen atom and highly negative localized negative charge on the fluorine, oxygen or nitrogen atom. This indicates that the electrostatic attractions between these molecules will be greater than the polar molecules that do not have hydrogen covalently bonded to fluorine, oxygen or nitrogen. For example,

- –O–H •••• O–H (e.g., water–water attraction)
- –N–H •••• O=C– (e.g., in a protein)
- –O–H •••• O=C (e.g., between water and sugar)

4.4   Bond energies

Single covalent bond ~ 300 kJ mol^{-1}
Hydrogen bonding ~ 100 kJ mol^{-1}
Dipole-dipole interactions ~ 50 kJ mol^{-1}
Van der Waals' forces << 50 kJ mol^{-1}

In order to get rid of the sizing material after the fabric manufacturing, low bond length in sizing material is desirable. Strong bonds like covalent and electrovalent create problems during desizing and enhance unavoidable desizing expenditure. Hence a judicious optimization has to be made between the adhesive power of sizing material and ease of desizing. A typical size paste dissolving strategy is shown in Fig. 4.2.

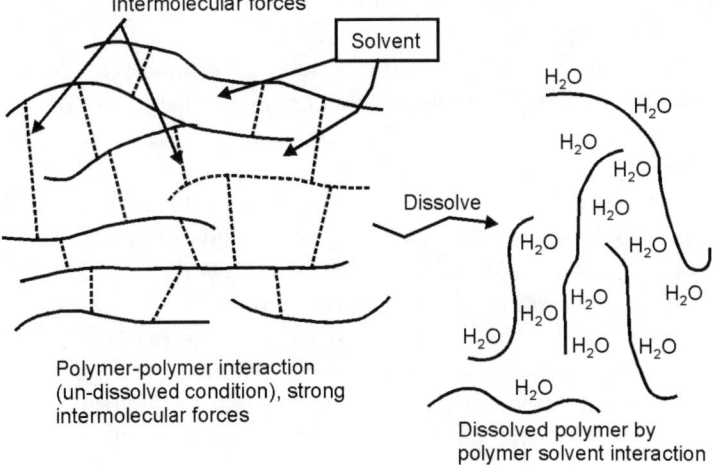

Fig. 4.2 Typical size paste dissolving strategy

4.5   Cotton yarn sizing with starch

The understanding of cotton fiber forming material, i.e. cellulose is essential to select an appropriate size material for it. The chemical structure of cellubiose (cellulose found in cotton) is shown in Fig. 4.3.

The bonding of chemical groups of the cellulose unit of cotton is shown in Fig. 4.3.

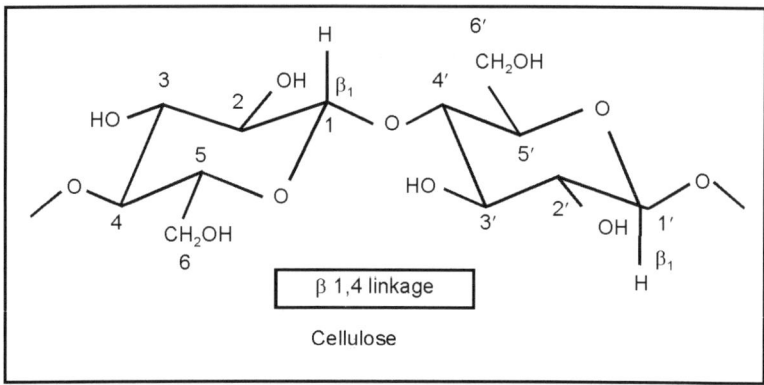

Fig. 4.3 Chemical structure of cellulose

It is evident from Fig. 4.3 that cotton (cellubiose unit) has a number of –OH groups that are potential groups for H-bonding. Hence, an ideal size material for cotton has to be one with the following characteristics:

- Large number of –OH groups
- High degree of polymerization
- Easy removal

After overall scanning of natural material available to do sizing of cotton yarns, a variety of polysaccharides are available for this purpose. Polysaccharides are cellulose, gums and starches.

Amongst these, cellulose (cotton linters and wood pulp) can't be used directly as it has to be treated and modified to make it soluble for application on cotton yarn, hence extra process and cost will add. Natural gums may be used, but their comparatively higher price restricts their usage. Hence, the above discussion leaves starch as the only alternative.

The repeat unit of starch (Fig.4.4) is usually composed of two components, a straight chain polysaccharide of glucose and a branched chain polysaccharide of glucose. Amylose (Fig. 4.5), the straight chain component, is relatively low in molecular weight, water soluble, and makes up to 20–30% of starch. Amylopectin (Fig. 4.6), the branched chain component, is relatively high in molecular weight, water insoluble, and makes up 70–80% of starch.

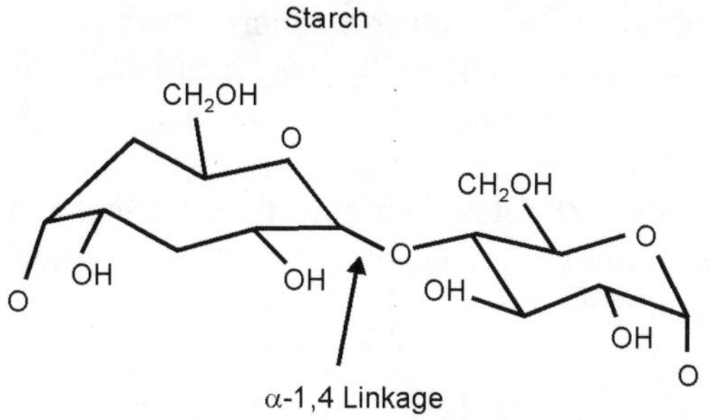

Starch

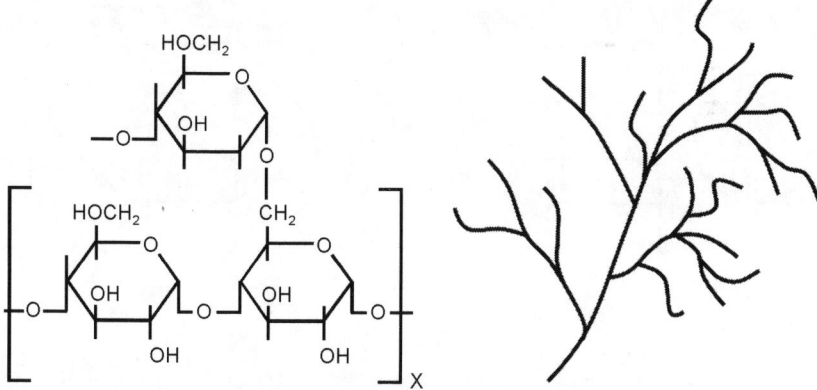

Fig. 4.4 Starch

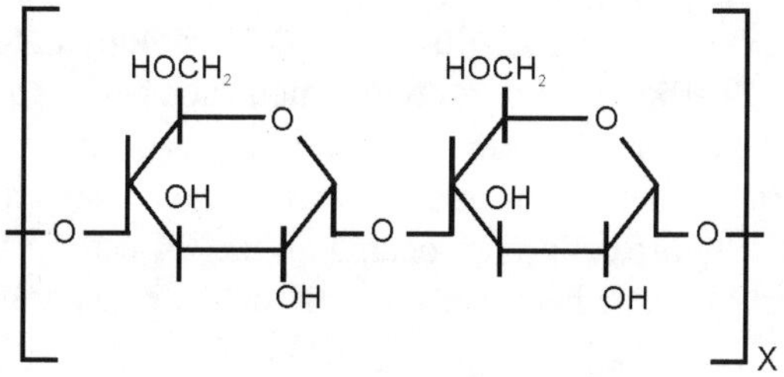

Fig. 4.5 Amylose

Fig. 4.6 Amylopectin

Starch remains available in granules form. The starch granules consist of both amorphous and crystalline regions. Starch granules are tightly bound and are unable to function as adhesive when suspended in cold water alone.

The granules consist of crystalline regions of straight chain molecules and straight chain sections of branched chain molecules, aligned. The crystalline regions are coupled with each other by more amorphous areas where the molecules are not aligned. Within the starch granule, the molecules and crystallites are arranged radically in concentric layers (Fig. 4.6).

Hence, these granules must be opened first to obtain adhesiveness. The starch can be made water soluble by oxidation, boiling, and acid and alkali treatment. These processes basically break the inter-polymer hydrogen bonds and make starch soluble.

4.6   Chemical analysis of starch

Both cellulose and starch, polymers of glucose, are identical in chemical constitution. The empirical formula of glucose ($C_6H_{12}O_6$) can be rearranged as $C_6(H_2O)_6$ to explain that glucose is a typical carbohydrate. Glucose is a member of ringed-carbon sugar group in which each carbon atom is attached with water molecule. This is explained well in Fig. 4.7, given below.

The anomeric (structural) forms of glucose molecule can be decided by the location of the hydroxyl group located on the C1 carbon of the pyranose ring which is part of cyclic hemiacetal involving the ring oxygen as shown in Fig. 4.8.

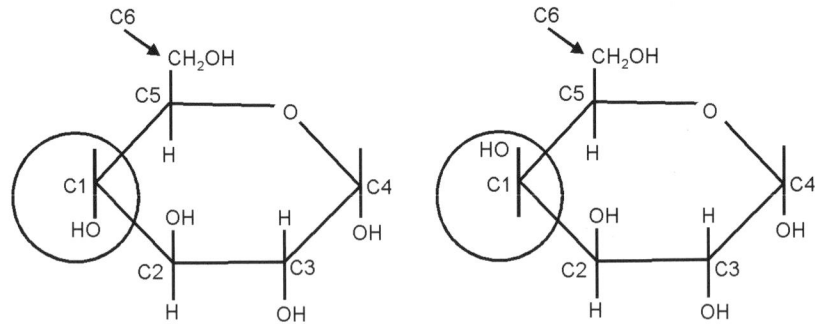

Fig. 4.7 α-D-glucose   **Fig. 4.8** α-D-galactose

It is possible for the hydroxyl group to be on C1 carbon to be above or below the ring through an interconversion with the ring oxygen, a phenomenon called mutarotation. The mutarotation offers two configurations of glucose called α-D-glucose and β-D-glucose.

Other hydroxyl groups of the pyranose ring, at C2, C3, and C4, are fixed at the up or down position. Actually, exchanging the position of hydrogen and hydroxyl groups gives a different sugar galactose as shown in Fig. 4.8. The two anomers of glucose are normally labeled α-D-glucose and β-D-glucose.

The condensation of glucose to form linear starch chains occurs between hydroxyl groups located at C1 and the C4 carbon atoms through elimination of a water molecule as shown in Fig. 4.9.

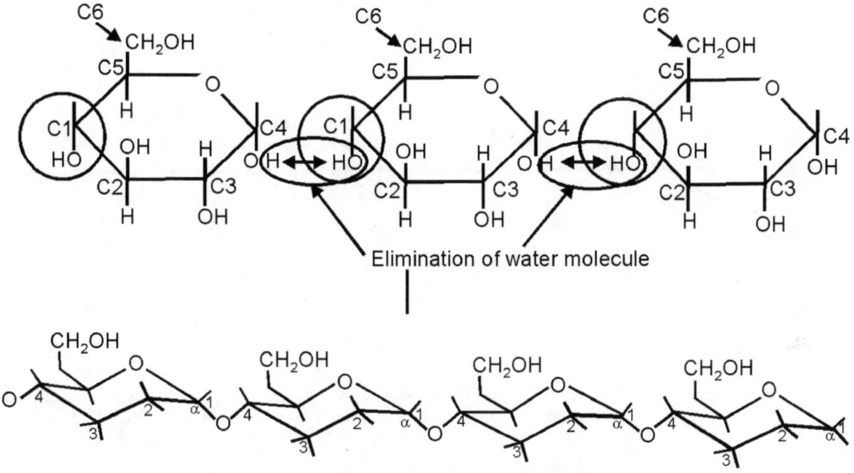

Fig. 4.9 Condensation of glucose to form linear starch

For synthesis of starch (α form of glucose), both of these hydroxyls are down with respect to the next glucose molecule. Hence, condensation of glucose units occurs without any change in the orientation of the glucose units. However, in the synthesis of cellulose (i.e., β form of glucose), the hydroxyl group is up at C1 and down at C4 carbon atoms. Therefore, every successive glucose unit must flip over before the hydroxyl groups at C1 and C4 carbon can match up for condensation, as shown in Fig. 4.10.

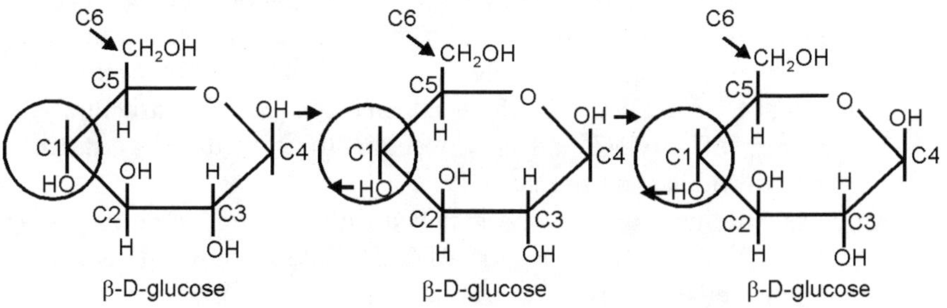

Fig. 4.10 Condensation of glucose to form cellulose

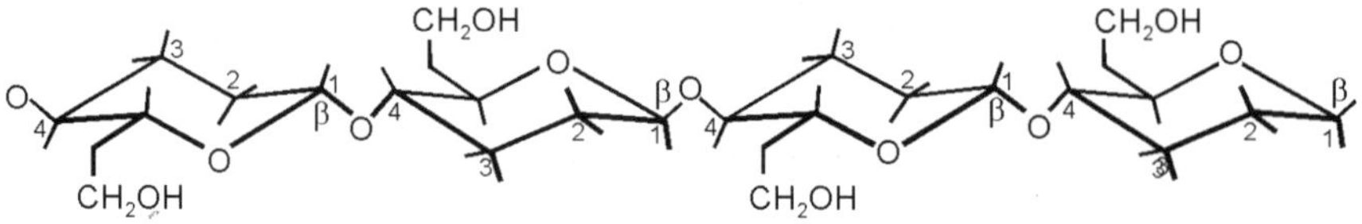

Fig. 4.11 Chain structure of cellulose (cellubiose form)

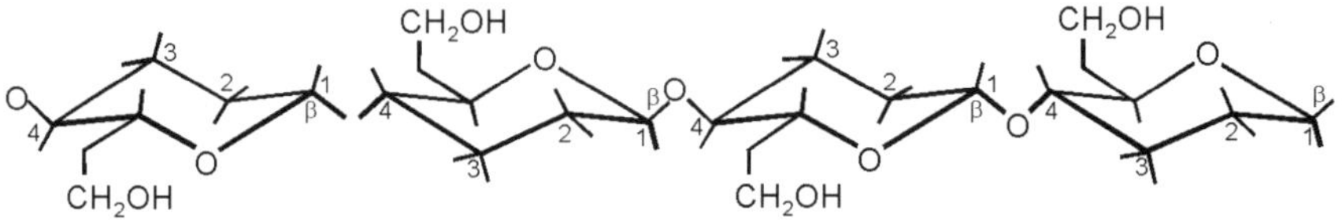

Chain structure of cellulose

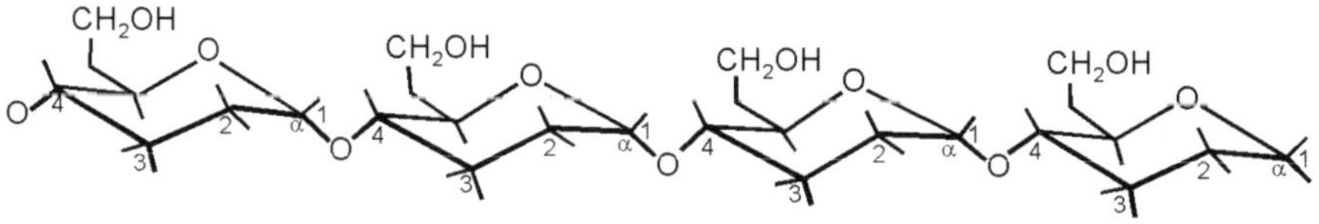

Chain structure of starch

Fig. 4.12 Comparison between chain structure of cellulose and starch

The comparison between linear chain structure of cellulose and starch is shown in Fig. 4.12. After the completion of condensation to enlarge the molecular chain of cellulose or starch, the inter-conversion possibilities at C1 carbon atom nullify and freeze the structure. Figure 4.12 reflects that both cellulose and starch have remarkable physical and chemical similarities. These similarities exhibit that both materials will have significant adhesion potential for each other. In consequence of that, a wide range of starches are used as sizing material to size different cellulosic fibers like cotton, viscose, and cellulose acetate, etc. On the basis of configuration, starches can be classified into two classes. The linear starch-molecular chain is called amylase, while highly branched chain structure of starch is called amylopectin. The condensation at C1 and C6 hydroxyls results in branched chain structure.

The content of amylase and amylopectin differs in different starches produced from various plant sources.

Amylose is a linear or sparsely branched carbohydrate based on $\alpha(1-4)$ bonds with a molecular weight of 105–106. The chains show spiral shaped single or double helixes. Amylopectin (Fig. 4.6) is a highly multiple-branched polymer with a high molecular weight of 107–109. It is based on $\alpha(1-4)$ bonds, and also on $a(1-6)$ links constituting branching points occurring at every 22–70 glucose units (Zobel 1988).

The process ability of different starches as size ingredients depends on the content of amylase and amylopectin. The role of the content of amylopectin is very crucial, because the presence of amylopectin retards the gelling which is micro-crystallization of molecular chains which helps to prevent the gelling of size paste at lower temperatures and enhances the process ability. The linearity of molecular chain of amylase is suitable for rapid crystallization and contributes to the strength and stability of size film on yarn surface.

Starches contain high amount of amylase and are advised to be cooked at a temperature higher than 150°C for better dissolution. It gels (retrogrades) quickly, if cooled. Genetic engineers and plant breeders have developed plant variants that produce starches approaching virtually 100% of either amylose or amylopectin chain structures. French (1984) revealed that the starch granules obtained from various plants are indeed sphero-crystals with limited crystalline regions. These crystalline regions are obtained by compaction of two types of chain structures into stratified layers formed around a nucleus. Imberty et al. (1991) concluded that the amylose and amylopectin chains are compressed into a parallel array of double helices with the 1–6 branch points being located in the amorphous regions as conceptually shown in Figs. 4.5 and 4.6. The starch can also absorb or desorb water depending upon the ambient temperature and relative humidity similar to cellulose. Pure and unmodified starch is known as pearl starch. The cooking process is used to feed the required thermal energy to the starch granules to permit water molecules to penetrate the compacted stratified layers of starch granules as shown in Fig. 4.15. The moment starch granules receive ample thermal energy at a particular temperature to overcome the potential energy of hydrogen bonding within the starch granules, gelatinization takes place. Schoch et al. (1954) revealed that as the gelatinization of starch granules takes place, crystalline region within starch granule disappears due to breaking of hydrogen bonds, and in consequence of that birefringence of starch granule disappears. Hence, the gelatinization temperature of starch granules depends upon the content of amylose and amylopectin, and coupling between these within the starch granule interior.

4.7   Starch cooking

Starch granules cooking process starts by adding appropriate amount of water with starch granules.

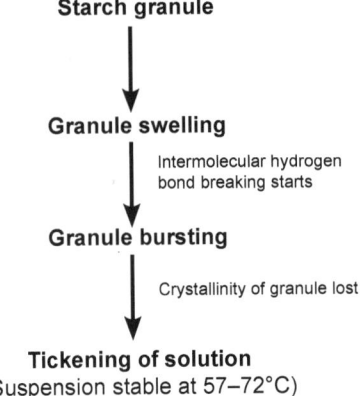

Starch granule

Granule swelling

Intermolecular hydrogen
bond breaking starts

Granule bursting

Crystallinity of granule lost

Tickening of solution
(Suspension stable at 57–72°C)

4.7.1   Kinetics of starch cooking process

Starch is found in granular form and boiling is the simplest method of breaking up starch granules. The kinetics of starch granules cooking process is shown in Fig. 4.13. During boiling, the starch granules first swell and then burst with coincident thickening of suspension. The temperature at which this thickening of suspension occurs is called gelation temperature, and in pure water this temperature is between 57°C and 72°C for starches. As water enters inside the starch granule, the intermolecular hydrogen bonds are broken. During gelation, the crystallinity of the starch solution is lost.

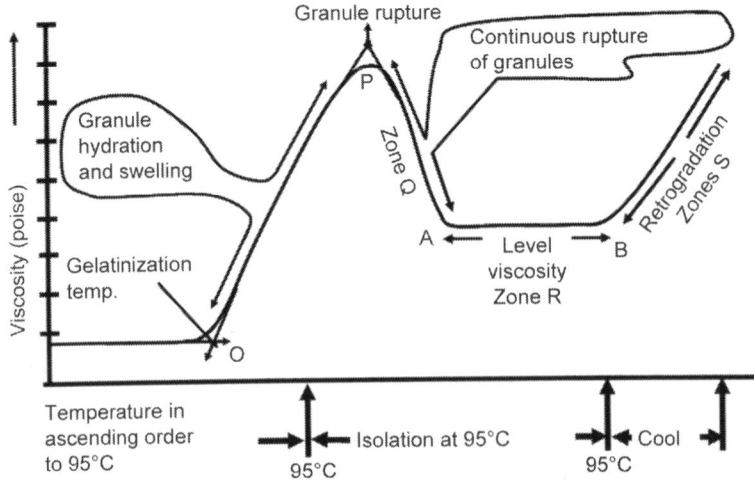

Kinetics of a typical starch cooking process

Fig. 4.13 Kinetics of starch cooking process

The kinetics of the starch-cooking process governs viscosity of the solution. The viscosity of the solution increases as most of the available water is absorbed by the continuously swelling granule at gelatinization temperature O. As the temperature of solution increases beyond the gelatinization point, the thermal energy of the solution super seat the intermolecular bonding resisting hydration of the amylose and amylopectin chains. As granule swelling remain continued and temperature reaches to granule-rupture temperature, the viscosity of the mixture continues to increase to a maximal level up to point P. Ultimately, the super-swollen granule, under the influence of the sheer force applied during cooking and stirring, bursts like a balloon, and swollen amylose and amylopectin chains disperse into size mixture. In consequence of that, a decrement in size viscosity takes place as shown by Q zone in Fig. 4.13, as concluded by Hall and Sayre (1971).

Continuous stirring and heating of size solution finally levels the viscosity after the rupture of all granules, called as level viscosity and shown by the zone R. The size paste is best suitable for coating the yarn surfaces. After relieving the kinetic energy from the cooked size paste and as the solution is cooled, the possibilities for multiple points for intermolecular chain contacts occur as shown in Fig. 4.13, and a rigid interlocked chain structure composed of a microcrystalline, micelle like structure having hydrogen bonding networks is formed, and the starch mixture gels as shown by a zone in Fig. 4.13. As the size is applied in this gel form at yarn surfaces, the gel dries and forms a film around the yarn surface.

The kinetics of starch-cooking process is explained in Fig. 4.13. As the temperature of size recipe mainly containing water and starch granules reaches up to 95°C, gelatinization takes place (till point O in Fig. 4.13) followed by granule hydration and swelling from point O posses significant hike in viscosity. When temperature reaches P, the granule rupture begins and fall in viscosity takes place that continue in zone Q. The size from temperature A to B attains a level viscosity and further rise in temperature results in retro gradation (zone S).

4.8   Crystalline structure of starch granule

Starch is composed of thin lamellar domains of about 4.5 nm thickness. Each lamella is made up of about 100 double-stranded helices, each consisting of about 20 glucose units. The double helices are very densely packed, with a high degree of regularity which forms the crystalline region of starch granules. The lamellar and superhelix structures of amylopectin are only a small part of the starch granule morphology. The starch granules are made up of alternating amorphous and *semi-crystalline* shells, between 100 and 800 nm thickness. These structures are termed growth rings. These growth rings consist both amorphous and crystalline lamellas. The amorphous part of the starch granules is studied least.

Cornuéjols (2009) explained that within the semi-crystalline regions of starch granules, the nanoscopic lamellae are parallel to the surface of the starch granule.

Starch granule structure is a nanoscale structure that is formed by double helices forming lamellae and the growth rings (alternating amorphous and semi-crystalline shells).

Fig. 4.14 Micelle-like structure in starch granule

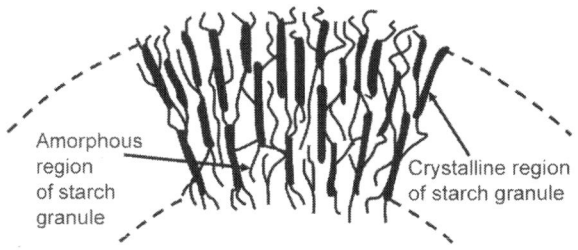

Amorphous region of starch granule

Crystalline region of starch granule

Fig. 4.15 Radial arrangement of starch crystallites

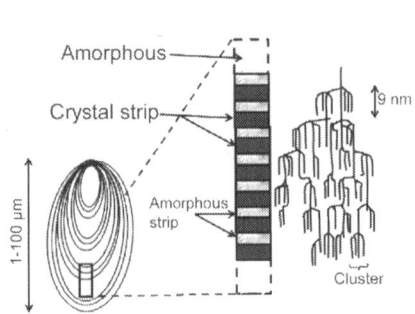

Amorphous

Crystal strip

Amorphous strip

9 nm

Cluster

1-100 µm

Fig. 4.16 Crystalline regions

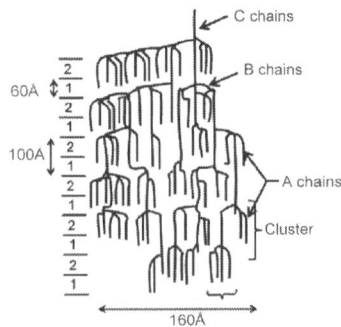

C chains

B chains

A chains

Cluster

60Å

100Å

160Å

Fig. 4.17 Starch granular structure

The disruption of granular starch is based on the transformation of semi-crystalline granule into a homogeneous, rather amorphous material with the destruction of hydrogen bonds between the macromolecules. Disruption of starch granules can be achieved by casting (e.g., with dry drums) or by applying thermo mechanical energy in a continuous process. The combination of thermal and mechanical stresses can be obtained by extrusion process. Extrusion process can be in one or two stages. In a one-stage process, the extruder, usually a twin-screw extruder is fed with native starch.

Starch shows a special granular structure (Gallant et al. 1986), containing a high degree of radial organization as shown in Fig. 4.15. Starch macromolecules are mainly oriented according to the radial axis. The ultra-structure is obtained by inter-macromolecules hydrogen bonding between hydroxyl groups with the participation of water molecules. Amylose and the branching regions of amylopectin form the amorphous zone in the starch granule. Amylopectin is the dominating crystalline component in native starch with double helix organizations. Habidi and Lucia (2012) revealed that the co-crystallization is also possible between amylose and free fatty acids or lipids. Several types of cristallinity are observed in starch granules in which A, B, and C types are prominent. Figure 4.16 shows that the crystalline regions (20–45%) are arranged as thin lamellar domains, perpendicular to the radial axis. Starch granular structure is a combination of A-chains, B-chains, and C-chains as shown in Fig. 4.17.

4.9   Reactive positions on the molecular chains of starch

Glucose residues of starch molecules decide its chemical reactivity. The chemical structure and reactive sites of amylose and amylopectin are shown in Figs. 4.18 and 4.19 respectively. Every glucose unit consists of six carbon atoms (referred as C1, C2, C3 . . . , etc.). Generally, there are two terminal hydroxyl groups that are found at the ends of the glucose molecular chains at C1 and C4. In case of each glucose residue of amylose, two secondary hydroxyl groups at C2 and C3, as well as one primary hydroxyl group at C6 are found.

In amylopectin, each branch point decreases the number of primary hydroxyl groups by one and simultaneously increases the number of secondary hydroxyl groups at C4 by one. All available hydroxyl groups of amylose and amylopectin are capable of being oxidized and reduced, as well as capable of participating in the formation of hydrogen bonds. These hydroxyl groups may play an active role to form salts and participate in the formation of ethers and esters. The starch esters may be formed with inorganic, as well as organic acids. The macrostructure of starch also plays a crucial role in its reactivity. The inter- and intra-molecular hydrogen bonding decrease the solubility of the material and restrict the accessibility of potential reaction sites to the reagents used, as discussed by Gotlieb (2005). The entanglement of amylase and amylopectin chains may also be a factor to decide the reactivity of starch. According to Cui (2005), these all factors do not have significant impact on reactivity of starch but may have noticeable impact. Hence, the selection of starch for sizing purposes requires underlined attention. The selection of the reagents and reaction conditions also requires special consideration. The granular character of native starch depending on penetrating ability of the reagent also influences the reactivity of starch. The polar character of the granule surface and the capillaries between granules prove starch a good porous substrate for absorption point of view. Various reagents undergo inclusion in the capillaries, forming starch capillary complexes.

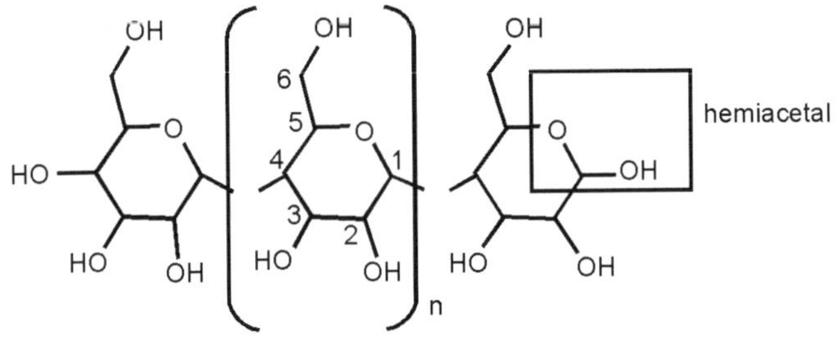

Fig. 4.18 Chemical structure of amylose

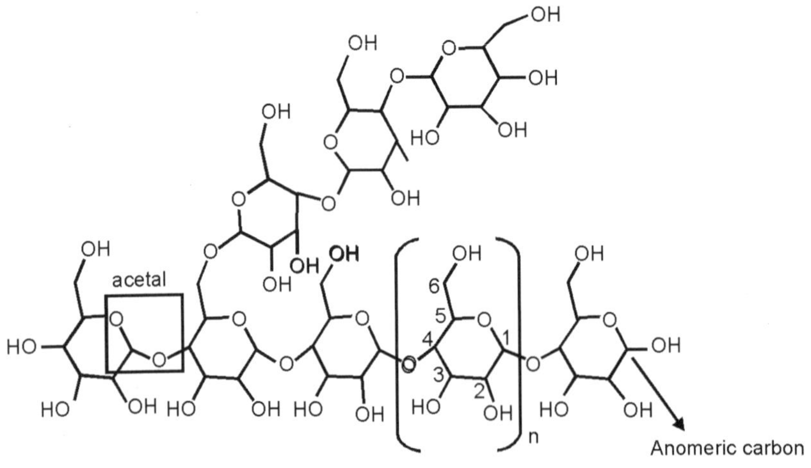

Fig. 4.19 Chemical structure of amylopectin

4.10   Modified starch

Modified starch, also called starch derivatives, are prepared by physically, enzymatically, or chemically treating native starch, thereby changing the properties of the starch. Starches are modified to enhance their performance in different applications. Modified starches are the main size materials used for yarn sizing of all natural fibers such as cotton, and linen, etc. For exploring the compatibility between starch and synthetic fiber, and flexibility of size film after starch modification in the sizing process of blended yarn such as polyester or cotton instead of poly vinyl alcohol (PVA) can be proved as a useful tool. Currently, starch derivatives have been the main green size material (environment friendly size material) due to its compatibility to synthetic fiber and flexibility of size film, as well as good stability and application at wide range of viscosity. When acetyl and hydrophobic ester group were induced into acetate starch, its adhesion to natural and synthetic fiber was enhanced. By introduction of ester group, the size adhesion, flexibility of size film, and strength elongation property has been increased.

Currently, acetate starch could substitute 20–30% PVA size worldwide in textile leading countries. In order to substitute more PVA, more attention has been paid to grafted starch by researchers. Grafted starch was prepared to be used in the sizing process of polyester or cotton blended yarns, and could substitute PVA in large scale. If graft technology was taken further, amelioration-grafted starch would be applied in large proportion, and even totally used instead of PVA. Polymethacrylic acid grafted starch prepared by Mostafa (1997) was obtained by grafting methacrylic acid to corn starch in potassium permanganate or citric acid system and performance of such grafted starch as size material was greatly enhanced. Fracture strength, breaking tensile rate, and weavability of such sizing yarn after cotton textile sizing were all excellent. Although there was a certain difference in size performance from PVA, it could be singly used in sizing process instead of PVA. Zhang et al. (2007) had prepared a grafted starch size with well permeability to fiber by grafting polyacrylic acid. Such size-possessed good sizing performance, size strength, and tensile rate, comfortable fuzzy yarns could be used in sizing process of polyester or cotton completely instead of PVA. However, how to totally substitute PVA in other textiles still needs further investigation (Qin 2006).

Starches may be modified to increase their stability against excessive heat, acid, shear, time, cooling, or freezing; to change their texture; to decrease or increase their viscosity; to lengthen or shorten gelatinization time; or to increase their visco-stability. Various modified starches are listed below:

• Dextrin roasted starch

- Acid treated starch
- Alkaline treated starch
- Bleached starch
- Oxidized starch
- Enzyme-treated starch
- Monostarch phosphate
- Distarch phosphate
- Phosphated distarch phosphate
- Acetylated distarch phosphate
- Starch acetate
- Acetylated distarch adipate
- Hydroxypropyl starch
- Hydroxypropyl distarch phosphate
- Starch sodium octenylsuccinate

Starch can be made water soluble by the following modifications:
(a) Acid treatment (thin-boiling starch)
Thin-boiling starches are made by adding a small amount of acid to starch suspension that is held just below its gel point.

Acid-modified starch is the new product obtained by soaking starch in the water slurry containing dilute mineral acid below the gelatinization temperature. The acid concentration must be sufficient to hydrolyze few bonds connecting glucose units in the starch chains within few hours. After neutralization, the product is recovered by filtering and drying to a condition indistinguishable from the starting material. The degree of modification of starch is governed by the amount of acid and time of reaction (Tharanathan 2005). Acid cuts the starch molecules in smaller size. Acid modified starch posses low viscosity opaque gel. The acid cleaves the polymer at the glycoside linkage thereby lowering the viscosity of a solution made from it as shown in Fig. 4.20. Hydrolysis occurs within the granule without breaking it.

Corn- and waxy-maize starches are the principal commercial parent starches used for this modification; though small amounts of tapioca, wheat, and potato starches are also used for acid modifications. Acid modified starches are produced in a wide range of fluidities. Acid-modified starch will characteristically have a lower hot-paste viscosity compared to the starting material at identical solids concentration.

The rheological properties of hot pastes of acid-modified starch are quite different than those of unmodified corn starch. In unmodified starches, the swollen granules have a prominent role in imparting thixotropic characteristics, whereas in acid modified starches, the granules have been damaged to a point where they disintegrate or fragment rather than swelling (Koch and Roper

1998; Yin et al. 1988). As a result, hot pastes of acid-converted starches are relatively fluid; however on cooling they retrograde, lose their clarity, and form opaque gels.

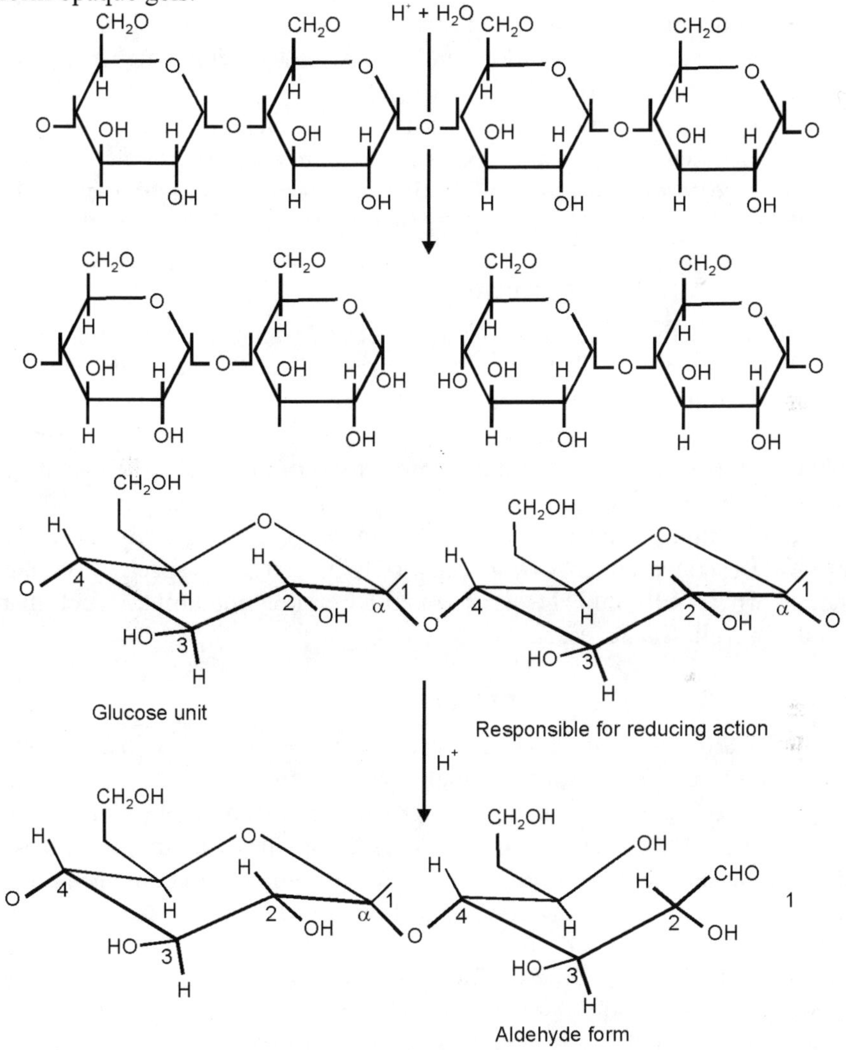

Fig. 4.20 Acid treatment of starch

Acid-modified starches are used in warp sizing of cotton and cellulosic based fibers as well as some cotton or synthetic blends. Different fluidities allow selection of the desirable starch for the type of yarn encountered. Coarse two-ply yarn requires little sizing for weaving and a thin-boiling starch of 20–

30 fluidity is adequate. Fine yarns need warp sizing with a 40–60 fluidity size paste. Normally, 10–15% concentrated size solution with softener/lubricant and preservative is used. Pickup can be hanged by varying the viscosity, concentration, machine speed, and squeeze pressure. Higher pickup from an acceptable viscosity-size solution can be obtained by using a higher fluidity starch at a greater concentration.

Acid modification plays an important role in the manufacturing of the other types of modified starches. Acid modification may be used as a pre-modification step in some cases, and post modification step in others. In both cases, acid treatment helps in getting a range of fluidities for the modified starch.

4.11   Alkali treatment

Strong bases activate the starch. Although starch granules can be completely gelatinized in aqueous alkali, the degree of granule swelling depends upon the nature of the starch, nature of the alkali, the relative amounts of starch, alkali and water, temperature, and presence or absence of neutral salts. In aqueous alkaline slurries, the starch granules absorb most of the alkali. The increased reactivity due to alkali absorption is a major factor in the manufacturing of starch derivatives commercially. A further increase in the reactivity of starch in aqueous alkali is obtained by the addition of neutral salts, especially sodium sulfate. These salts shift the starch-alkali absorption equilibrium, such that alkali absorption is increased.

4.12   Oxidation of starch

Commercially oxidized starch is manufactured by reaction of an alkaline aqueous suspension of starch granules with hypochlorite solution, which converts the 2–3 hydroxyls into –COOH groups, breaking the ring at that point (Fig. 4.21). Generally, 5–7 –COOH groups per 100 anhydroglucose are introduced. Sodium bisulfate is added to destroy excess hypochlorite. The granular structure is retained, and films from oxidized starch perform better than acid-modified starch. The oxidation causes de-polymerization which results in lower viscosity of dispersion and introduces carbonyl and carboxyl groups. Oxidized starches are produced in a wide range of fluidities. In general, the fluidity rises with the increase in the level of oxidation.

After oxidization, hydroxyl group changes to ketone, aldehyde, or carboxyl groups by cutting the starch molecular chain at different places. The paste of oxidized starch is low in viscosity and stretchable in wet condition. Oxidized starch provides highly stable paste after cooling also.

The introduction of oxidized starches is a distinct improvement over unmodified and acid-modified starches for many purposes. Oxidized starches have a number of properties which make them attractive as textile sizes. The starch granules cook at a lower temperature so that a uniform viscosity is reached sooner. The paste of oxidized starches can be prepared at higher solids than unmodified starch, and the pastes when allowed to cool will be distinctly more fluid, less gel, and clearer than unmodified starch or acid-modified starch pastes. This behavior is attributed primarily to the presence of carboxyl group introduced into the starch molecule during oxidation, which sterically interferes with the tendency of amylose to associate and retrogradate. Oxidized starch forms tough, clear, and continuous films. They tend to give more homogeneous films which have fewer tendencies to shrink and crack than the films of unmodified or acid-modified starches. Their films are also more soluble in water due to the hydrophilic nature of the carboxyl group introduced during oxidation.

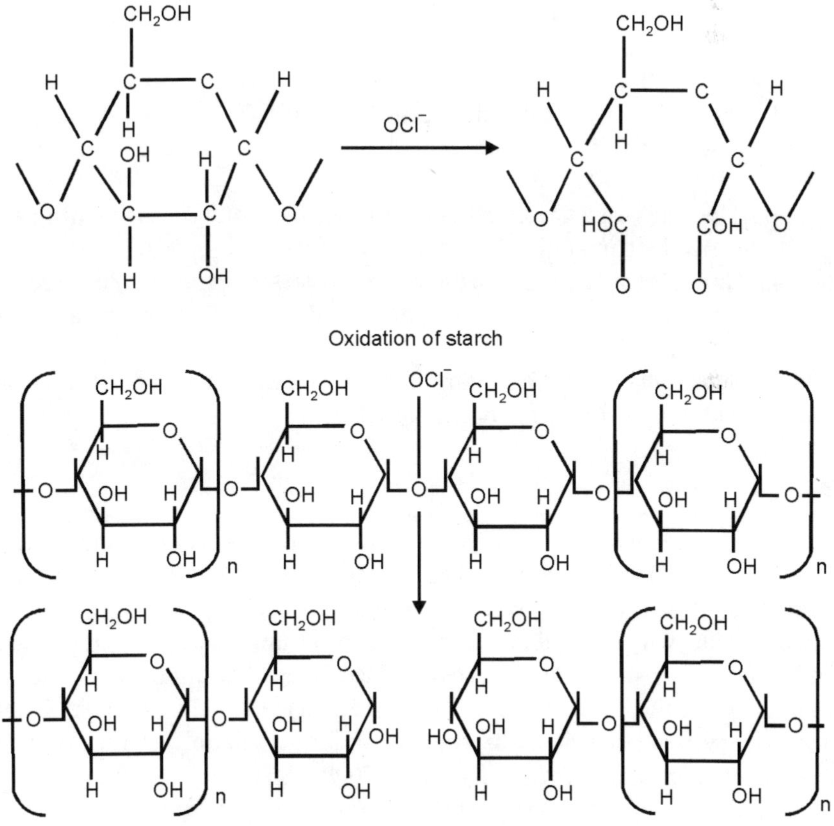

Oxidation of starch

Fig. 4.21 Oxidation of starch

Hypochlorite oxidized starches play an important role as size or getting the above mentioned properties. However, with the introduction of newer starch derivatives, e.g. ethers and esters, the use of hypochlorite oxidized starches are diminished.

Thin boiling starches when blended with starch, gave considerably better properties. Blended yarn such as P/C and P/V can indeed be sized with thin boiling starches at significantly higher concentration than the sizing of cotton.

4.13   Substituted starches

Modification by substitution involves the replacement of the hydrogen atom of hydroxyl group with the mono-functional group to produce starch ethers, esters, and phosphates. The comparison of substituted starch with unmodified starch is as under

1. Substitution of relatively smaller number of hydrogen of hydroxyl groups into the starch molecule reduces the gelatinization temperature by 10–30°C compared to unmodified starch.
2. Presence of bulky substituent prevents the secondary bonding association of adjacent hydroxyl groups and subsequent formation of a hard size.
3. The films of substituted starches can be removed easily from the yarns than unmodified starch.
4. Due to reduction in re-association, pastes of more stable viscosity are obtained on cooling or aging, and facilitate their use at lower temperatures.
5. Substituted starch form films that are uniform, soft and flexible; hence addition of softener is not essential.

4.13.1   Starch esters

Low degree of substitution (DS) starch acetate (Fig. 4.22) is commercially available since 1950 in plenty. Starch acetates are synthesized by treating suspension of starch with acetic anhydride under slightly alkaline conditions. Starch acetate when heated in water shows lower initial swelling temperature and a stable viscosity paste with a reduced tendency of gelation. Starch acetate films are more flexible, elastic, and transparent than unmodified starch. Starch acetates are used extensively in warp sizing for cotton, polyester/cotton, and other synthetic fiber blends (Xiao et al. 2009). Starch acetate films can be desized easily without enzyme and has good adhesion to cotton and other blends.

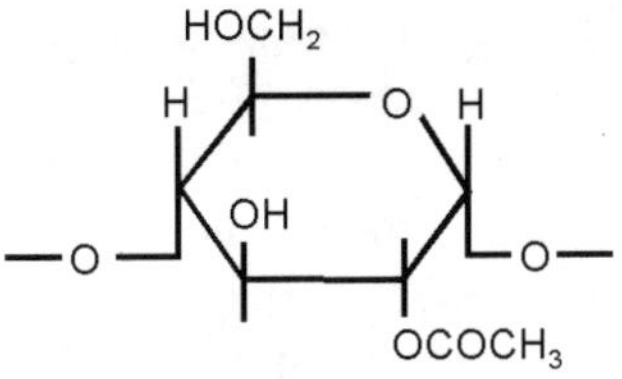

Fig. 4.22 Starch acetate

Another sizing composition suitable for sizing of synthetic yarns is prepared by acetylating starch with dibasic acid anhydride and forming salts of starch with amines (devoid of acidic hydrogen) at pH 5–7. Potato-starch esters have been used as a universal size for all types of yarns. Starch esters can be used as a sizing agent for cotton, wool, and nylon yarn. The increased penetration by the soluble product is thought to improve the size-fiber bond. Hence, the weaving performance of the yarn sized with starch esters is improved.

4.13.2   Starch ethers

Ether linkages among the organic compounds are known to be the most stable, and starch ethers are also one of them. Amongst the wide varieties of starch, ethers carboxymethyl starch (CMS), hydroxyethyl starch (HES), and cationic starch are commercially used for sizing of textiles (Shruren 2009).

(i) *Carboxymethyl starch (CMS)*
Carboxymethyl starch (CMS) is a derivative of cellulose synthesized by its reaction with alkali and chloroacetic acid. Reaction of starch with sodium-chloroacetate in aqueous alkaline slurry produces a low substituted carboxymethyl starch. Recovery of the product by filtration and washing takes place in original granule form.

At low concentrations, CMC molecules are extended (rod-like) mostly but at higher concentrations, the molecules overlap and coil up. Hence at high concentrations, the molecules entangle to become a thermo-reversible gel. Both increasing ionic strength and reducing pH decrease the viscosity as they cause the polymer to become more coiled.

Generally, the degree of substitution can be varied up to three and can be produced in a wide range of viscosities. CMS acts as an anionic polyelectrolyte in the aqueous slurry. This derivative has a lower initial gelatinization temperature, increased rate of swelling, and higher viscosity than the unmodified starch (Ragheb 1988 and Xiao et al. 2009). The chemical structure of the carboxymethyl starch is shown in Fig. 4.23. The pastes are cohesive, have increased clarity, and the tendency to gel is greatly diminished. The films are flexible, clear, and have improved solubility in water. Schultz prepared the most widely used carboxymethyl starches such as

- High-viscosity derivative having a DS 0.29
- Low-viscosity derivative having DS 0.28

Rheological behavior of carboxymethyl starch, cross-linked carboxymethyl starch gels, and benzylated carboxymethyl starch ethers were studied. Partial benzylation (DS between 0.15 and 0.30) of CMS is used for sizing of cellulosic materials; sticking of paper (Mazíková et al. 2009).

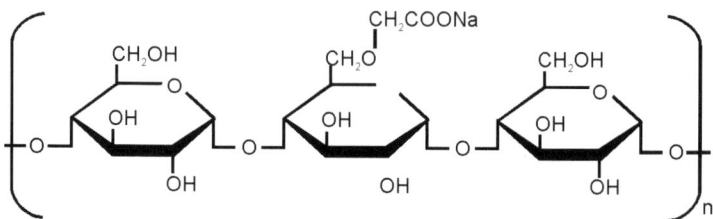

Fig. 4.23 Carboxymethyl starch

Most CMCs dissolve quickly in cold water. CMCs are mainly used for controlling viscosity without gelling. It may be used to improve the volume yield during baking by encouraging gas-bubble formation because its viscosity drops during heating. Its control of viscosity allows its use as thickener, phase, and emulsion stabilizer (for example, with milk casein), and suspending agent. CMC can also be used for its water-holding. Thus, it is used for retarding staling and reducing fat uptake into fried foods.

(ii) *Hydroxyethyl starch (HES)*

One of the most common representatives of the commercially successful starch ethers is the hydroxyethyl starch ether made by the reaction of alkaline suspension of starch with ethylene oxide. Substitution is kept low up to 0.1 molar substitution (MS) because of significant changes in the physical properties of the derivative. Starch is reacted and recovered in its native form and the principal effect of substitution is found during the pasting and film formation. The hydroxyethyl group is nonionic, hence they are not affected by electrolytes and pH compared to the ionic starches. By substituting a small number of hydroxyethyl groups into the starch granule, following changes take place:

- The starch pastes at a lower temperature
- Granules swell faster and more uniformly
- Pastes are clearer
- Possesses more cohesiveness with warp yarn surfaces
- After cooling, paste will not set to a firm gel

Hydroxyethyl starch is considered to be one of the best film-forming starches. The substitution of starch with hydroxyethyl group produces upgrading effects on the film characteristics if the substitution is limited up to 0.1 MS.

Hydroxyethyl starch is used to size cotton and cotton-blend warps including polyester, acrylics, and nylon spun yarns. Low substituted hydroxyethyl starches are used alone or in combination with PVA in the warp sizing. Hydroxyethyl starch ether is used as a sizing agent for polyester cotton-blended yarn and it is concluded that the product can replace a part of PVA in sizing recipes.

Hydroxyalkyl starch ether cross linked with epichlorohydrin gives products which are able form solutions of better stability, moisture resistant films, and improves size to yarn adhesion (Zobel and Stephen 2006).

(iii) *Cross-linked starch*

Cross-linked starches are produced from the reaction of starch with bi-functional reagents to introduce cross-bonds between the glucose units of the separate chains. Such reagents include epichlorohydrin, sodium trimeta-phosphate, phosphorous oxychloride, formaldehyde or other polyfunctional reagents. Minor amounts of cross linking will have a significant effect on pasting and gel properties of the starch. Initial swelling and gelatinization temperature is raised and the reaction can be controlled to yield products of long-term stable viscosity than untreated starch. Cross-linking plays a very important role when supplemented to modified starches, such as thin-boiling and substituted starch, etc. Cross-linked acid modified starch with sodium trimetaphosphate is then treated with malefic anhydride at a pH 8–10; this product was more suitable for P/C blend yarns. Potato, corn or wheat starch was cross-linked with 0.15–0.2% epichlorohydrin in presence of dilute sodium hydroxide and then oxidized with sodium hypochlorite to give the cross-linked oxidized starch, a product containing one crosslink and 10–14 carboxyl groups per 200 glucose units (Abraham et al. 2007). Figure 4.24 shows the structure of cross-linked starch.

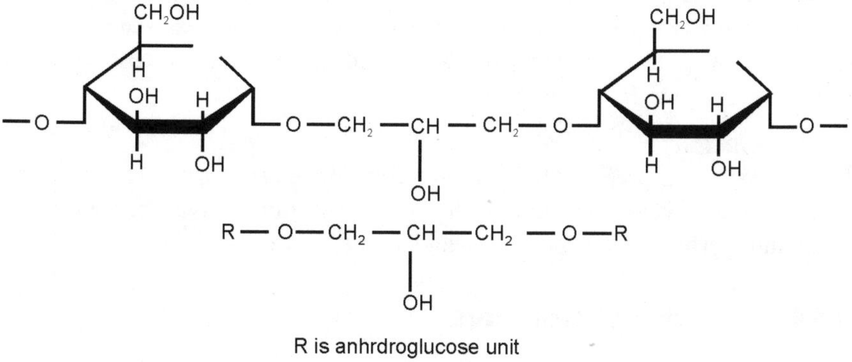

R is anhrdroglucose unit

Fig. 4.24 Cross-linked starch

(iv) *Grafted starch*

There has been considerable interest during the past years in the block and graft copolymers of starch (Victoria et al. 2005). Most of the work is based on the initiation of free radical on starch and then allowing the radical to react with polymerizable vinyl and acrylic monomers, thereby producing a polymer chain of improved functionalities. Graft copolymers exhibit the physical and chemical properties of both the main chain polymer and the copolymer chain grafted on it.

Monomers that give water-soluble starch-graft polymers are acrylic acid, acrylamide, and acrylate ester containing quaternary ammonium groups. Water-insoluble starch grafts are obtained with acrylonitrile, alkyl acrylate esters, and styrene (Moftosa 1995).

Saponified starch-g-polyvinyl acetate has been patented as a sizing agent for cotton, rayon, and polyester yarns. Oxidized starch grafted with 2-hydroxyethyl acrylate and acrylic acid gave good yarn to size adhesion compared to unmodified starch. Oxidized starch grafted with partially saponified vinyl acetate gave graft copolymer having better film properties. A blend of monoesters of a polyvalent fatty acid anhydride with oxidized-g-starch (with a hydroxymethyl acrylate) was used as a sizing aid for polyester yarn. Yarn to size adhesion was good with this copolymer compared to the yarn sized with polyvinyl alcohol.

Sizing agents used for sizing cotton or polyester/cotton warps were prepared by grafting starch with an unsaturated ethylenic compound containing carboxyl group or its derivatives; optionally hydrolysing the polymer for changing the viscosity. Thus, 400 parts of starch were grafted with 50 parts of acrylic acid to give polymer paste. Polyester/cotton warps sized with (10% solids) the above copolymer paste gave a weaving efficiency of 98.3% as compared with 87.9% for warps sized with saponified polyvinyl acetate.

(v) *Dextrin-roasted starch*

Dextrins are a group of low-molecular-weight carbohydrates produced by the hydrolysis of starch. White and yellow dextrins from starch roasted with little or no acid is called British gum.

(vi) Bleached starch

Bleached starch is prepared by treating starch with a very light oxidation using various acids such as sodium hypochlorite, sodium chlorite, hydrogen peroxide, potassium permanganate, peracetic acid, and ammonium persulphate.

4.14   Synthetic binders

With the introduction of synthetic fibers and their blends, the process of sizing became very challenging. The important factors like penetration of size, cohesion and adhesion of the size with the yarn, film elasticity, and frictional properties, etc. play a leading role towards the performance of the sized synthetic fibers and their blend warp yarns in weaving.

Synthetic fibers have high tenacity, low moisture regain, and static charge generating tendency. Though the main objective of sizing cotton is to minimize loss in elongation-at-break, increase strength and ability to resist abrasion, but the problems are different for synthetic fibers. Synthetic yarns are strong and do not require improvement in strength. But being hydrophobic, it is difficult to make adequate adhesion of sizing material with synthetic fibers. Synthetic fibers blended with cotton become more hairy than cotton, and require strong, elastic, and flexible films with better adherence.

All these factors necessitate the use of a sizing material different from traditional use for cotton yarns. The selected sizing chemicals must modify the mentioned above properties to the synthetic blended yarn in order to improve the weaving performance. Carboxymethyl cellulose (CMC), polyvinyl alcohol (PVA), and polyacrylates have been introduced as sizing adhesives.

4.14.1   Cellulose-based binder

Carboxymethyl cellulose (CMC) or sodium carboxymethyl cellulose (Fig. 4.25) is water soluble cellulose based binder. It is made by reacting monochloroacetic acid or its sodium salt with alkaline cellulose and reagent properties, temperature, and concentration are major process parameters. CMC is soluble in hot and cold water. The solutions prepared from pure CMC are neutral and stable in 3.5–11 pH range. Apart from easy solubility, CMC has many other advantages as mentioned under:

- Micro-organisms do not attack the paste or the film of CMC. Therefore, there is no need to add an antiseptic.
- The paste has not only good storage properties, but is also slow congealing.
- The viscosity changes are small over a wide range of temperature.
- The film is uniform and soft; therefore, the use of a softener can be eliminated or minimized.
- It has excellent adhesive and binding properties.
- Low or medium viscosity grade CMC have excellent binding properties.
- The film is water soluble. Therefore desizing is easy.
- CMC has better adhesion for acetate rayon or synthetic fibers than starch or soluble starch.
- It is used extensively along with starch or soluble starch for the blend yarns.

- CMC has been found to have excellent use in sizing for all types of fibers.
- Its blend with starch is useful for cotton warps.
- It is useful alone for continuous filament viscose rayon or acetate rayon warps.

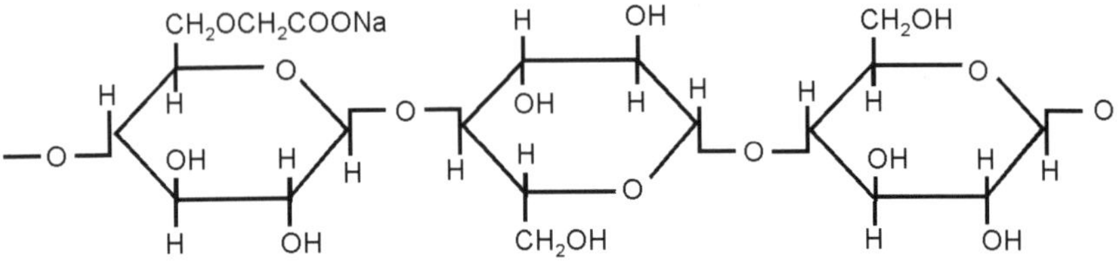

Fig. 4.25 Carboxymethyl cellulose (CMC)

4.14.2 Vinyl-based binders

Vinyl-based binders include polyvinyl alcohol and acrylic polymers.

(i) *Polyvinyl alcohol (PVA)*
Polyvinyl alcohol (Fig. 4.26) is a film-forming polymeric substance having secondary alcoholic groups on alternate carbon atoms of aliphatic micro molecules. Generally, PVA is prepared by hydrolyzing polyvinyl acetate (Fig. 4.28) by replacing acetyl group of the vinyl radical by hydroxyl group. The PVA powder is soluble in hot or cold water depending upon the extent of hydrolysis.

There are two main varieties of PVA:
1. *Fully hydrolyzed (FH) PVA* – All the acetyl groups are hydrolyzed in (FH) PVA
2. *Partially hydrolyzed (PH) PVA* – Significant part of about 7–20% or more content remains unhydrolyzed.

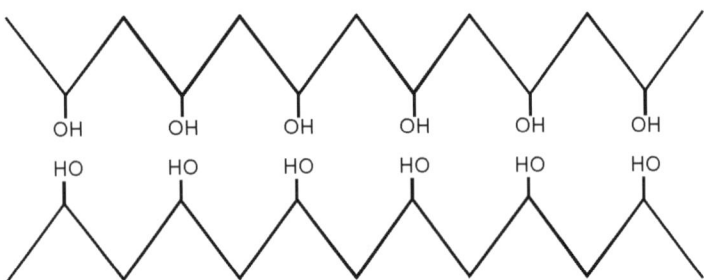

Fig. 4.26 Fully hydrolyzed polyvinyl alcohol (PVA)

Modi studied the properties of PH and FH grades of PVA and suggested the necessary properties while incorporating in a PVA size recipe (Modi 1983). The comparable data reveal that FH grade PVA is more suitable than PH grade PVA for sizing of blended yarns.

The polyvinyl alcohol (PVA) paste has good storage property due to its non-congealing tendency. It requires antiseptic in small quantity to be added only if several months' storage is essential. The film of PVA has less moisture regain and more strength than that of CMC.

Polyvinyl alcohol (Fig. 4.27) as an adhesive in sizing has many merits. It gives good penetration inside the yarn. Therefore, it serves as a good binder if used as an addition to starch. It has very good adhesion properties, both in natural and manmade fibers. Its film has good abrasion resistance and tensile strength (Harper 2006).

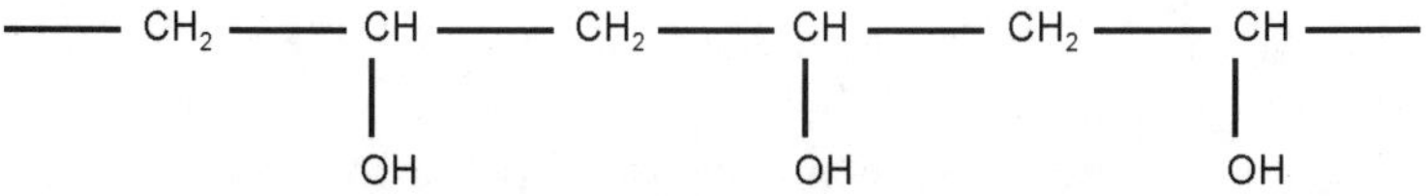

$$\text{---CH}_2\text{---CH---CH}_2\text{---CH---CH}_2\text{---CH---}$$
$$\text{OH} \qquad\qquad \text{OH} \qquad\qquad \text{OH}$$

Fig. 4.27 Polyvinyl alcohol (PVA)

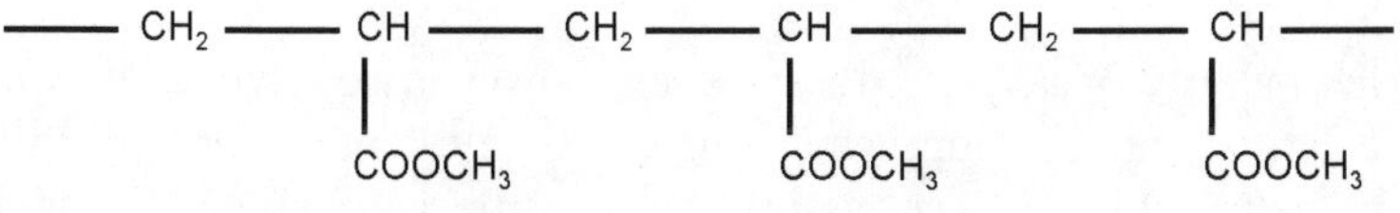

$$\text{---CH}_2\text{---CH---CH}_2\text{---CH---CH}_2\text{---CH---}$$
$$\text{COOCH}_3 \qquad \text{COOCH}_3 \qquad \text{COOCH}_3$$

Fig. 4.28 Polyvinyl acetate (PVA)

PVA can be used on all fibers, but its use on cotton warps is rare because of its high price. It is frequently used for manmade fiber warps like rayon, synthetics, and their blends. PH grade exhibits greater adhesion to hydrophobic fibers such as polyester, and enhance the abrasion resistance of the polyester rich sized yarn warps made of staple fibers and their blends. The fully hydrolyzed grade is preferred along with other adhesive like CMC and starch, etc.

(ii) *Acrylic polymers*

With the development of synthetic fibers, increased diversification of fibers and improvement of quality of fabrics, demands for acrylic sizing agents are being made. Acrylic sizing agent is the general name of homopolymer, copolymer or their combination of acrylic monomers. Synthetic size with the main body of acrylic ester is the paramount kind of acrylic sizing agents, and has excellent adhesion to hydrophobic fibers like polyester, tender size film with lower strength, large deformability, and moisture absorption stickiness viscosity. After sizing with acrylic polymers, the number of fuzzy points on yarn surface is reduced.

The blame of environmental issues on PVA size opened the new venues for investigation and development of acrylic sizing agents.

- Poly acrylate, polyacrylamide, and polymethyl acrylate are the three major types of acrylic size in which poly acrylate type has excellent adhesion to hydrophilic fiber, large moisture absorption, heavy stickiness viscosity, and would have well sizing performance in combination with low content of modified starch in sizing recipe.
- Polyacrylamide (PAAM) type has large moisture absorption and heavy stickiness viscosity, and good adhesion to hydrophilic fiber, with 25% of solid content. It is a kind of monomer homopolymer, and its quality is easy to control. It could be made to solid formulation.
- Acrylate type is the high technological product of acrylic acids, and a copolymer of acrylate monomers, with two or three monomers. It has good adhesion to hydrophobic fiber, and modified moisture absorption stickiness viscosity as compared to the previous two types.

It is possible that combination of acrylic size and modified starch was applied in the sizing process of thin cotton fabric completely instead of PVA (Qin 2006).

The majority of acrylic adhesives are, however, copolymers of various acrylic derivatives. The copolymerising units are esters of acrylic acid (Fig. 4.29) or methacrylic acid, acrylamide , acrylonitrile, and sodium salts of either acrylic acid or methacrylic acid. Recently, there had been a spate of research in the field of acrylic polymers in sizing. The advantages of acrylic size are as follows:

- High tensile properties of the film
- High adhesive power for synthetic yarns
- Best results at low relative humidity
- Good abrasion resistance of the film-fiber matrix
- Easy desizing
- Consistent viscosity with the optional distribution of the adhesive within and on the surface of the yarn

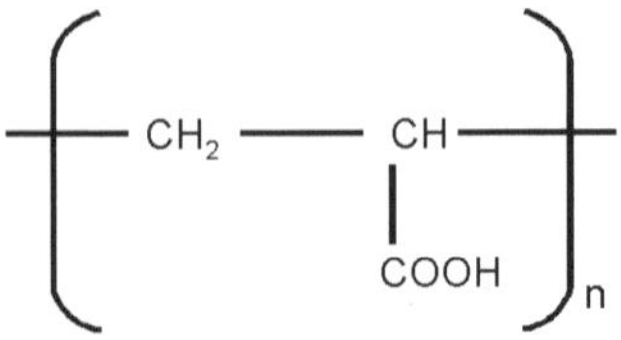

Fig. 4.29 Polyacrylic acid 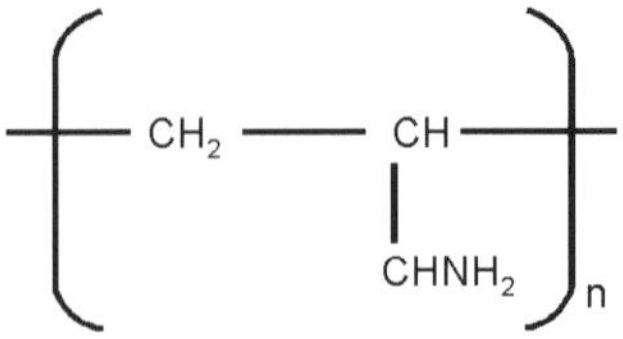   **Fig. 4.30** Polyacrylamide

A common disadvantage of all acrylic polymers is their plasticity and loss of tensile strength at high relative humidity (more than 70% relative humidity). Acrylic sizes can be engineered to satisfy the end usage requirements as size ingredients. On the basis of the properties of monomer, copolymer can be synthesized. New age acrylic sizing agents are architected to make a balance combination of hydrophobic/hydrophilic characteristic suited for various blend yarns are being produced. It is proved by various researches that the acid component imparts adhesion and solubility and its ester gives elasticity and improves adhesion while the acrylonitrile/acrylamide component improves the toughness of the film and the compatibility of the size to hard water and heavy metal ions. Copolymer involving acrylic ester/methacrylic acid has been tried or water jet size. A copolymer of ethyl acrylic acid and methacrylic acid (sodium salt) has been found suitable for sizing polyester, nylon and acrylic fiber yarns (Athar 1983).

A new type of acrylic size to obtain higher adhesion to polyester/cotton in order to elevate size film strength, flexibility, and fuzzy yarns comfort with commercial name "STSX-1" developed by Donghua University was prepared by copolymerization of hydrophilic monomers, i.e. acrylamide and acrylic acid, and hydrophobic monomers, i.e. butyl acrylate and vinyl acetate. The size film of STSX-1 had lower fracture strength and lower leasing resistance. Moisture absorption of its size film is similar to PVA, with no thickness viscosity, lower cost and good stability. By using STSX-1 consumption of PVA in the size-paste recipe was reduced from 55% to 25%, and less environmental pollution was imposed.

In general, the acrylic ester products are required for sizing of polyesters, whilst the acrylic acid based products lend themselves to the sizing of polyamides and acrylic acid salt based products are most suited to cellulosic fibers.

4.15   The behavior of starch solution

The variation of viscosity of starch solution with temperature is shown in Fig. 4.31 by Brabender viscosity graph. Initially, as water molecules enter the starch granule, the starch swells up and then bursts. This is done by an increase in the viscosity of the solution. On further heating, more interpolymer H-bonds break. This is shown by decrease in viscosity of the solution. Finally when the solution is cooled, there is again formation of H-bonds as the paste thickens, and this process is known as retrogradation of starch.

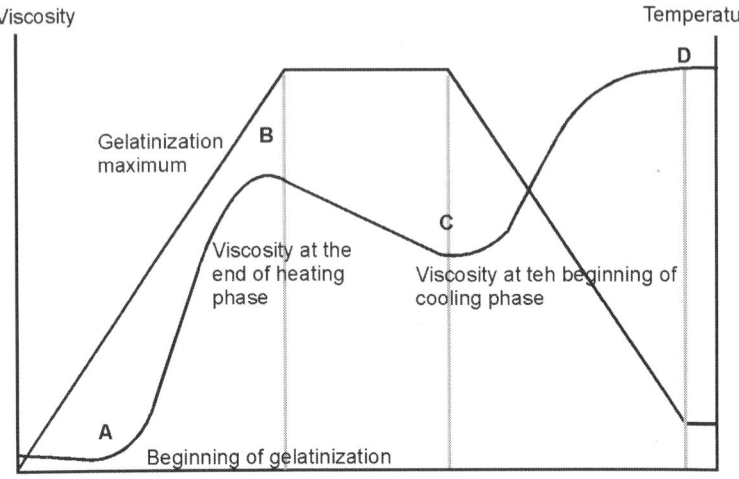

Fig. 4.31 Brabender Viscosity graph

4.16   Sizing of synthetic yarns

Synthetic yarns behave differently during sizing and weaving, and pose a very different problem in further processes. Following reasons are basically responsible to raise problems in synthetic yarn sizing.

- Absence of –OH groups
- Hydrophobic nature

Polyester and nylon fibers are the major representatives of synthetic yarn group, whose structures are shown below.

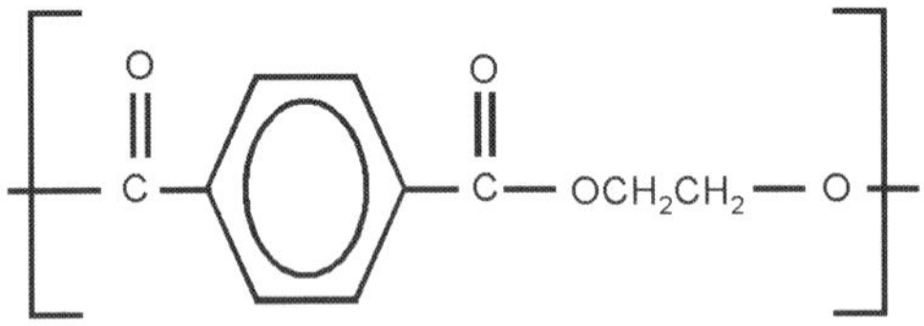

Fig. 4.32 Repeat unit of polyester

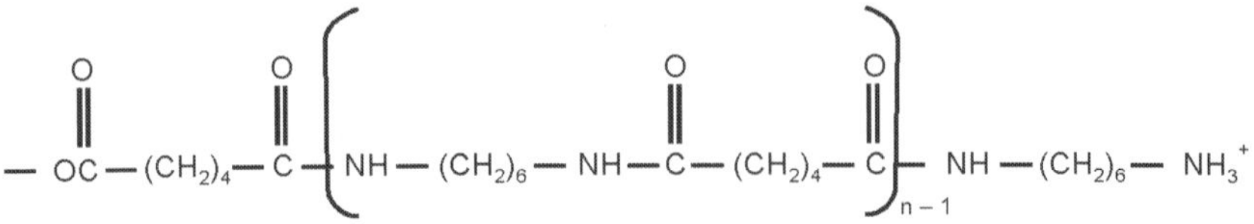

Fig. 4.33 Repeat unit of Nylon 6,6 (a typical polyamide)

It is evident from Figs. 4.32 and 4.33 that there are no –OH groups, that are mainly responsible for H-bonding and moisture absorption. Both PET and nylon consists C=O group that is also capable of forming the H-bonds, but the ease of formation of H-bonds is less than that of C-OH group. This can be attributed to the fact that it is easier for the O atom to draw electrons across a single bond rather than double bond. It is evident from nylon 6, 6 structure that very few H bonds can be formed (Fig. 4.34).

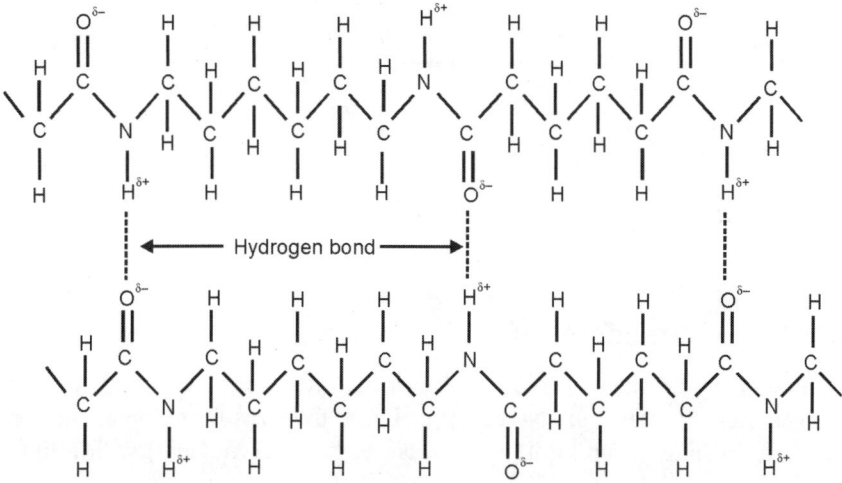

Fig. 4.34 Hydrogen bonds in nylon 6,6

Hence for sizing of these hydrophobic yarns, we shall have to primarily depend on the following forces that will play a decisive role:

- Van der Waals' forces
- Dipole–dipole forces

Hence in order to size the yarns, following arrangements have to be made:

- Size ingredient must not contain steric hindering groups
- Size and fiber have to be brought in close vicinity
- Size molecules must be as much linear as possible
- Sizing material should have favorable dipole-dipole interacting (polar) groups

The sizing agents used for polyester, nylon, and acetate yarns are listed below:

4.17   Fiber size

Polyester vinyl copolymers
Nylon polyacrylic acid
Acetate polyvinyl alcohol

4.17.1   Size for polyester

Polyester fiber is hydrophobic in nature and hence to size polyester yarns basically on Van der Waals' forces and dipole-dipole forces, it will play major role in order to achieve the required bonding (as hydrogen bonding is not possible due to limited places for hydrogen bonds). Hence, the size material for polyester should have linear chain with an attractive group.

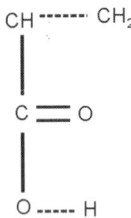

Fig. 4.35 Poly vinyl acetate

4.17.2   Size for nylon

Although, nylon fiber is hydrophobic in nature, it has the ability to form H-bonds as it contains the N-H group. Hence, the sizes for nylon contain the H-bond forming group (–OH in the case of poly acrylic acid) with a linear chain (Fig. 4.36).

CH ------ CH$_2$
|
C ═══ O
|
O ----- H

Fig..4.36 Acrylic acid

Langston (1952) made a series of experiments on sizing of spun and nylon yarns by using various sizing ingredients like dimethylol urea, polystyrene, starch, gluconic acid, polyvinyl alcohol, polyacrylic acid, polymethacrylic acid with added plasticizers or thickeners, polymethacrylic acid, and sodium alginate. Langston inferred that the most promising sizing ingredient for nylon appear to be olymethacrylic acid, polyacrylic acid, and polyvinyl alcohol. The sizing efficiency of all these materials is increased by the addition of 1–3% of sodium alginate to the size recipe. A size mixture containing 4.0%

of polymethacrylic acid and 1.8% of sodium alginate has been used and it has improved the weavability of 18/1 spun nylon warp.

4.17.3   Size for acetate rayon

Cellulose acetate rayon also has the H-bond forming group –OOCH3. Hence, a suitable hydrogen-bond forming size material group with a linear chain is suitable for this type of fiber. One such size material is poly vinyl alcohol (PVA) as shown in Fig. 4.28.

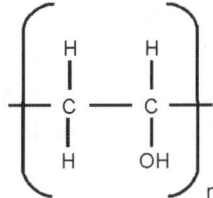

Fig. 4.37 Poly vinyl alcohol

Trommer (1957) did a pioneer work to launch a concept for identification of sizing ingredients on fabrics woven from synthetic filament yarns. The number and types of sizing ingredients used on filament synthetic yarns were constantly increasing, thereby requiring the commission dye house to constantly alert about the potential desizing difficulties that may accompany with each new sizing ingredient. Polystyrene resin, extensively used in sizing filament acetate warps, becomes insoluble when acidified and may precipitate out of solution and deposit on the cloth surface if not thoroughly rinsed after an alkaline scour, before giving any acid treatment. Hence an acetate rayon fabric, warp sized with polystyrene resin, never advised to give any acid treatment before proper desizing. Several other sizings are similarly precipitated by acid media. Polyacrylic acid becomes very tenaciously set on nylon if the fabric is heat-set before this sizing has been removed.

4.18   Hot melt adhesives

These are derivatives of ethylene vinyl copolymers, which are specially formulated for application without the use of water. These sizes are formulated in such a way that they can melt easily on application of heat, and also contain hydrogen bond forming (=O) groups, that increase with the bonding of the fiber. In case of ethylene vinyl terpolymers, the maleic anhydric group increases the adhesion of the size to the fiber. These sizes have the unique advantage that their adhesive property can be engineered as per the requirement, that is done by suitable selection of chain lengths of each group (Figs. 4.38 and 4.39).

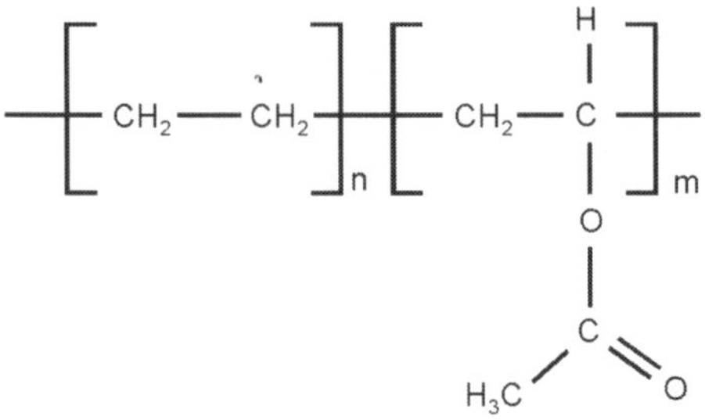

Fig. 4.38 Ethylene vinyl acetate

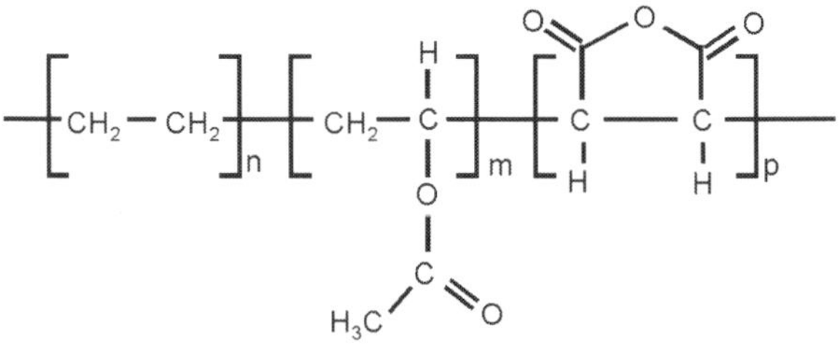

Fig. 4.39 Ethylene vinyl acetate terpolymers

A prime advantage of these sizing agents is their ability to flex easily during the weaving cycles while remaining adhered with the fiber. The other advantages of this size are given below:

- High cohesive strength and compatibility
- Excellent adhesion to a wide range of substrates
- High resistance to rupture

4.19   Analysis of sized yarn under the influence of squeezing pressure

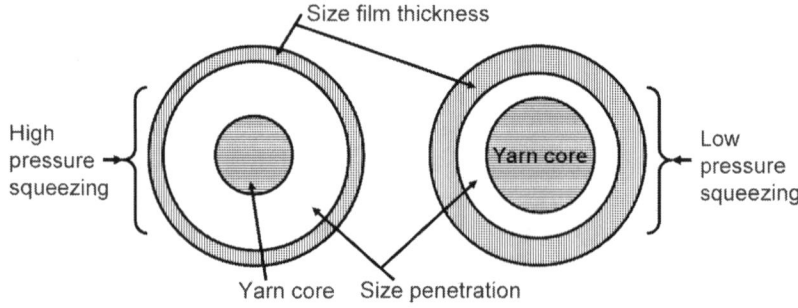

Fig. 4.40 Effect of squeezing pressure on size film thickness

4.20 Size pickup

Size pickup is the amount of size material picked up by warp yarns. In ideal condition, the size pickup must be pre-decided according to yarn or fabric construction because of the various process parameters like the formulation of size recipe, squeeze-roll pressure, level of size in size box, viscosity of size paste, and other mechanical conditions. Size must be picked up homogeneously by warp sheet.

4.21 The sizing machine

The material flow in a typical sizing machine is shown in Fig. 4.41 by means of a block diagram.

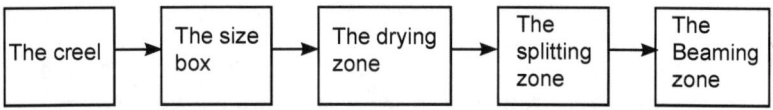

Fig. 4.41 Passage of material in a typical sizing machine

4.21.1 Creel

Creel in sizing machine is used to arrange warp beams in such a way that yarn tension among the all warp threads from different warp beams must be uniform.

Warp beams are arranged in a creel in such a manner as to facilitate

1. an easy withdrawal of warp sheet from each beam,
2. proper assembly of each sheet into the final form in which sizing should take place,
3. easy access of the operator to each element in the creel. Creels are designed for judicious utilization of the space and maintaining the tension uniformity in the threads of an individual beam during unwinding from full beam to empty state as also between the threads of the different beams as low as possible. Ideally, all the threads across the entire sheet entering the sow box must be entered in size box under same tension throughout the sizing process.

Double-deck beam stand sizing creel

A typical double-deck beam stand sizing creel is shown in Fig. 4.42. Total number of required beams is divided into two rows. One row of beams is mounted at a certain height higher than other. Threads from both rows are withdrawn and mixed at the middle of both the rows.

Double deck beam stand

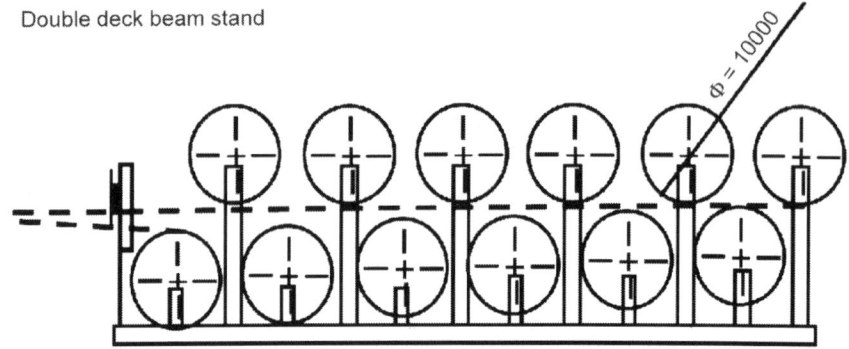

Fig. 4.42 Double-deck beam stand sizing creel

Four-beam section creel

The arrangement of four beams per unit would ensure easy access of the operator and better utilization of the vertical space but posses varying path length and wrap angle of the sheets from individual beams. A typical four-beam section creel is shown in Fig. 4.43.

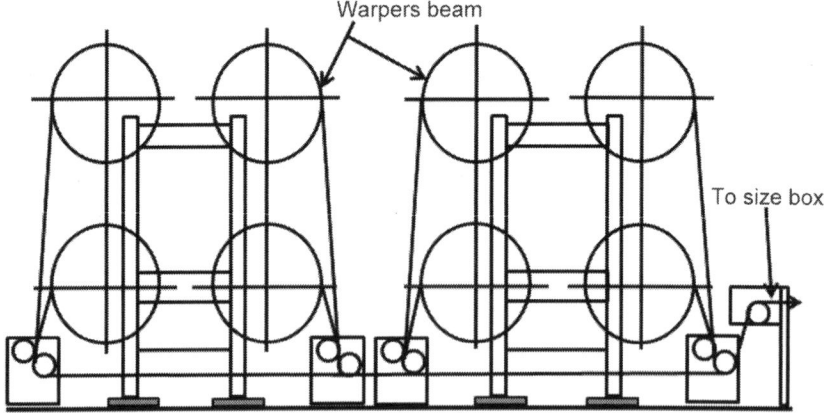

Fig. 4.43 Four-beam section sizing creel

Alternate up-down creel

The arrangement of the beams in an offset fashion is a distinct improvement on the former in terms of tension variation, but the access and space utilization is not as efficient. The third arrangement provides the optimum solution in terms of tension uniformity, strain, and accessibility. A typical alternative up–down creel is shown in Fig. 4.44.

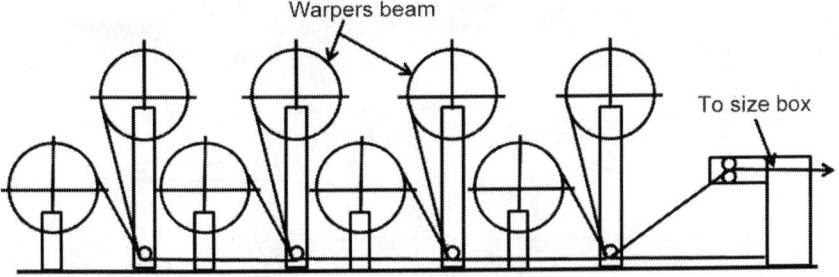

Fig. 4.44 Alternative up-down creel

Two-tier creel

Enhanced abrasion of the warp sheet coming from the top tier with beams of the lower tier may become the cause of significant diameter difference in between beams of upper and lower tier. A typical two-tier creel is shown in Fig. 4.45.

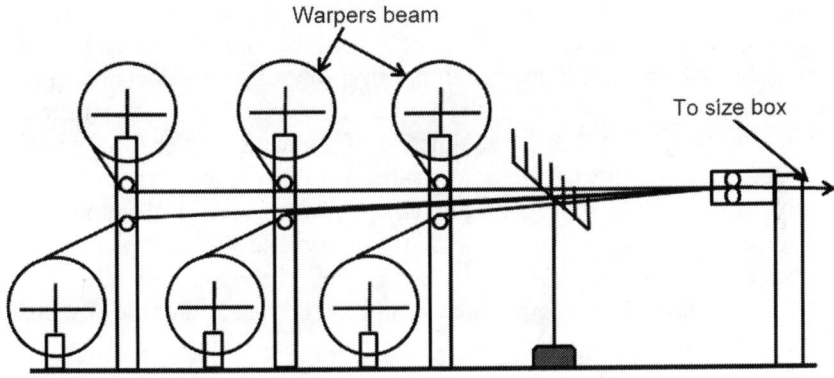

Fig. 4.45 Two-tier creel

Irrespective of the arrangement of the beams in a creel, the rotation of each beam is controlled by brake. A brake may typically consist of a rotor and multiple pneumatic operated calipers (up to a quantity of four calipers). The quantity of calipers is based on tension requirements. With this configuration, the brake-caliper pads are easily replaced, and cost effective.

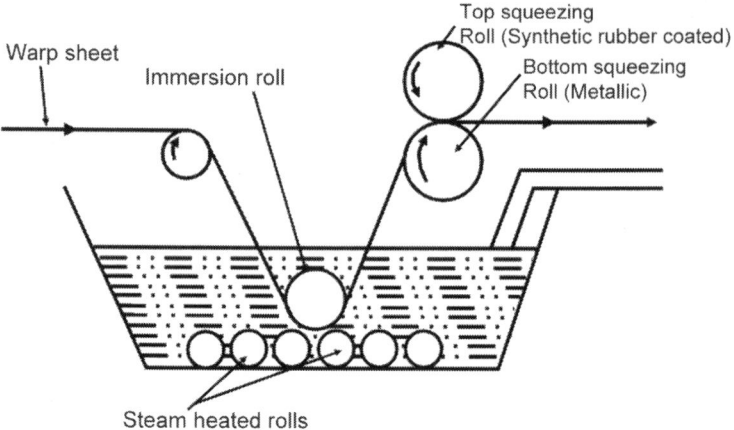

Fig. 4.46 Conventional type size box

4.21.2   The size box

The size box is the core part of any sizing machine and is situated ahead of the creel section. In case of a conventional size box, the assembled yarn sheet is taken to the size box and immersed in the size liquor by an immersion roller. The conventional size box consists of a guide roll to guide the sheet uniformly to the immersion roller. The prime function of immersion roller is to provide the uniform accessibility of size paste to all warp yarns in the warp sheet. The size box, shown in Fig. 4.46, is usually equipped with heating steam pipes.

The depth of immersion roller helps to regulate the size pick up by warp sheet. To maintain the viscosity and prevent the phenomenon of retrogradation, a series of steam pipes are equipped at bottom part of the size box. After immersion, the warp sheet reaches to a pair of squeezing rollers in which bottom roll is made of metal works as anvil, while the top roll is rubber coated to provide a cushion to warp sheet during squeezing against metallic bottom roller. The hardness of rubber-coated roller is a crucial parameter to decide the squeezing performance of sizing machine. Generally, the sore hardness of top squeezing roller is kept up to 50–70°.

To overcome all these shortcomings of size box, various sizing machine manufacturers have done different modifications in size box. Few of them are discussed here.

Modern size box
The size box, shown in Fig. 4.47, is usually equipped with heating coils linked to temperature control system for maintaining the temperature and therefore

the viscosity of the size liquor. The depth of the immersion roller below the level of the liquor is adjusted in conjunction with the speed at which the yarn sheet is taken through the box so that the net amount of liquor picked up by the yarn is kept at a desired value. Higher the speed of yarn, lesser would be the dwell time of yarn within the liquor for any specific setting of the immersion roller. Reduced dwell time would be expected to result in less-size liquor pick up if the viscosity of liquor is kept constant. On the other hand, higher speed of yarn in the liquor would also mean a higher drag force, proportional to the square of the yarn speed, which would result in a larger amount of liquor being carried away by the yarn. There is therefore equilibrium among the dwell time, viscosity, and pick up. It is important that the level of size liquor, which affects dwell time, be also held constant during the entire sizing process. This necessitates a level control system that maintains a steady inflow of size liquor to the box. An overflow system is illustrated in Fig. 4.48.

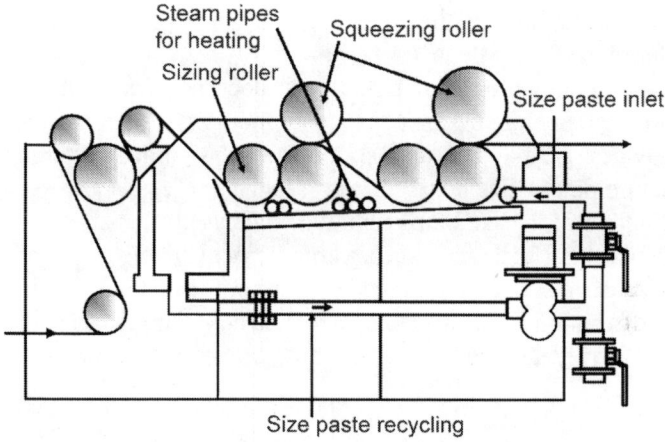

Fig. 4.47 Modern size box

The yarn sheet emerging from the size liquor is next taken through squeeze rollers where, some excess amount of liquor carried by the yarn sheet is squeezed out and flows back into the box.

- The effect of the nip of the squeeze rollers on performance of sized yarn is very crucial.
- Sore hardness and squeezing pressure are the final process parameters used to decide add-on on warp yarns. This conventional-type size box provides one-dip one-nip operating action. These conventional-type size boxes have various limitations that restrict both the productivity of machine and quality of sized yarn

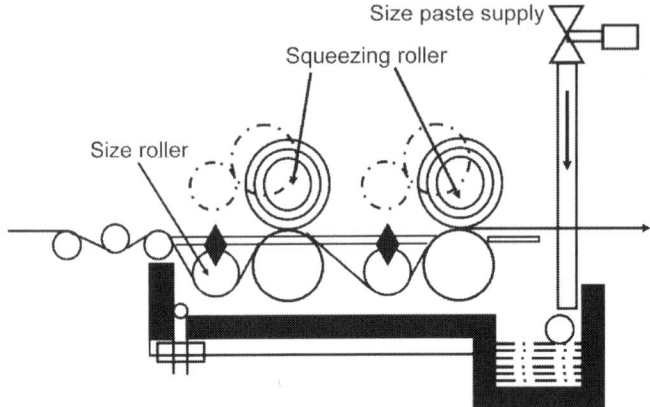

Fig. 4.48 Essential elements of a modern size box

- Very difficult to maintain the size liquor level constant for a long duration
- Temperature profile of size liquor does not remain homogeneous throughout the size box due to limitations of heating and stirring devices. Size liquor temperature varies significantly from wall to centre part of size box
- Number of dips and nips cannot be increased
- Pre-wetting arrangement for yarns is not available
- Size paste supply is arranged by very crude method which results in the variation in size paste level in size box, time to time

Effect of squeezing pressure on size film thickness is pictorially illustrated in Fig. 4.40.

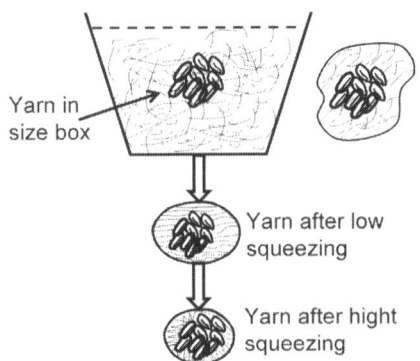

Fig. 4.49 Effect of squeezing pressure on sized yarn

During its passage through the nip, the wet yarn gets flattened causing a temporary reduction in the number of voids within the yarn body (Fig. 4.49). After coming out from the nip, the yarn tends to recover to its original form. During this process of recovery of shape and void, an inward radial flow of size liquor into the yarn body takes place. Thus a controlled squeezing results not only in the right amount of wet pick up, but also in the desired extent of penetration of size into the yarn body. Penetration of liquor to less extent within the yarn body would result in a surface coating, which can easily flake off while an excessive penetration would enhance the yarn stiffness. A couple of rollers, one with a hard surface and the other with a relatively compressible one (rubber coated), are usually employed for creating an effective nip of squeeze rollers.

Generally in squeezing roll pair, bottom roller made of steel and top of nitrile rubber-cover roller is preferred to generate a pressure in the range of 1–5 ton per linear meter. The hardness of the two surfaces, and also the smoothness and concentricity are the critical parameters to achieve a desired level of squeezing pressure on all the threads uniformly across the entire width of the warp sheet. A set up with two pairs of squeezing rollers and two immersion rollers is illustrated in Fig. 4.50. Such a double dip-double squeezing is more effective than single dip-single stage squeezing if the viscosity of liquor is high and/or the speed of sizing is high. A double squeezing can also be effected through an alternate arrangement of the squeeze rollers, depicted in Fig. 4.50. In such an arrangement, the compressible roller is located between two harder ones. The vertical arrangement would effectively be a single dip-double squeeze system, while the horizontal one would result in double dip-double squeeze. A special feature of the horizontal arrangement is the withdrawal of the sized sheet after the second squeeze. By avoiding any rubbing with walls of the squeeze roller after coming out from nip, the effectiveness of gluing of protruding fibers with yarn body is increased.

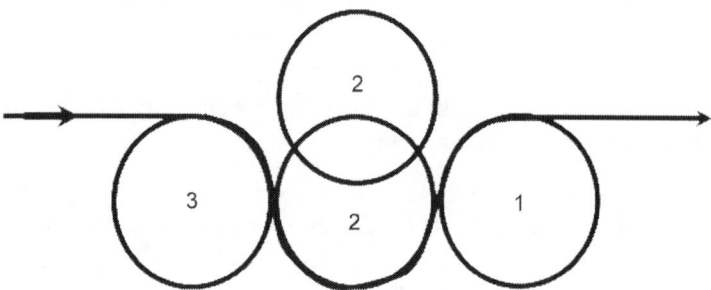

Fig. 4.50 Three roller squeezing system

4.22   Benninger squeezing system

Benninger developed a different type of squeezing system in which the inlaid glass bead on the roller surface is provided, as shown in Fig. 4.51. This bead provides a unique structure which is maintained throughout the roller life. This typically structured roller surface is designed to prevent the sticking of threads in order to reduce yarn hairiness.

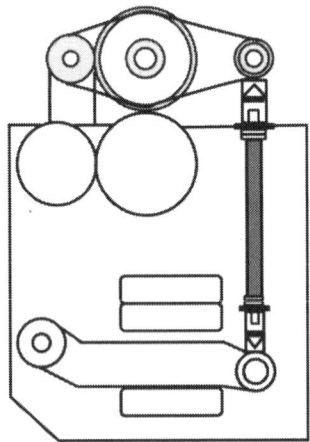

Fig. 4.51 Benninger squeezing system

4.22.1   Dry pick up

Dry pick up is the ratio of mass of bone-dry size material picked up to that of the corresponding mass of the bone-dry yarn expressed as percentage.

$$\text{Dry pick up} = \frac{\text{Bone dry weight of sized yarn} - \text{bone dry weight of unsized yarn}}{\text{Bone dry weight of unsized yarn}}$$

4.22.2   Wet pick up

Wet pick up is the ratio of mass of size paste picked up by the corresponding mass of yarn expressed as percentage.

Determination of size pick-up
The size pick-up can be calculated in different ways, and these can be classified in two groups
1. Continuous determination on the sizing machine
2. Lab tests of size pick-up on warp samples

The lab-test method can be further classified in two groups
1. By gravimetric method
2. By desizing according to ASTM-D334-GOT standard

Testing and analysis of size pick-up data are useful to set process parameters and take curative actions to control the end-breakage during weaving. Gravimetric method of size pick-up measurement is not standardized till date. The gravimetric method is used for possible controls according to sample drying prior to sizing to absolute dryness, weighing, air-conditioning, sizing, drying to absolute dryness and final weighing. Size pick-up may be determined by desizing according to the above standard.

Continuous determination of size pick-up

Continuous determination of size pick-up is possible by integrating a sensor-based device for measuring few parameters during sizing. These parameters must be controlled to maintain constant size pick-up during the sizing process.

The industrial humidity tester by Pleva GmbH Germany continually measures warp humidity after passing through the size box and size concentration in the box. Size pick-up is determined on the basis of these two parameters. Rest of the sizing parameters are considered constant in this procedure. This procedure enables to determine the size pick-up during sizing and to correct it by pressure alteration. This procedure offers the greatest advantage of determining the size pick-up.

However, this process did not consider the effect of change in warp moisture prior to sizing, which influences the size pick up as confirmed by some other researchers also[Kovac_ević, S. 2004; Kovac_ević et al. 2002; Orešković et al. 1975]. The size pick-up equation is based on the concept of substance balance.

$$S_p = \frac{W_{sp} - W_h}{\dfrac{100}{C} - \dfrac{W_{sp}}{100} - 1}$$

Where S_p is wet size pick-up, W_{sp} is warp moisture at the box exit (%), W_h is warp moisture at the box entry (%), and C is size concentration in the box (%).

4.22.3   Thin-boiling solution

Thin-boiling solutions are the solutions that have high concentration at much lower viscosity compared to the normal ones.

4.22.4   Relation between concentration, viscosity, dry and wet pick-up of normal and thin boiling solution

It is evident from Fig. 4.52 that at the same concentration k_2, the thin boiling starch solution has much lower viscosity η_1 as against η_2 of the normal starch solution. Similarly, if the viscosity of solution is kept same at the value of η_1, the concentration of the normal starch solution is k_1, a value lower than k_2, which is the corresponding concentration of the thin boiling starch. As a result, for the same wet pick-up, the dry pick-up with thin boiling starch is higher by the amount ΔP_1.

Conversely, for the same dry pick-up, less wet pick-up is required with thin-boiling starch, thus requiring less water to be evaporated from the sized yarn. Thus for same viscosity $\eta1$, the increase in dry pick-up with thin-boiling starch is ΔP_1, whereas for the same concentration of paste k_3, the decrease in dry pick up is ΔP_2.

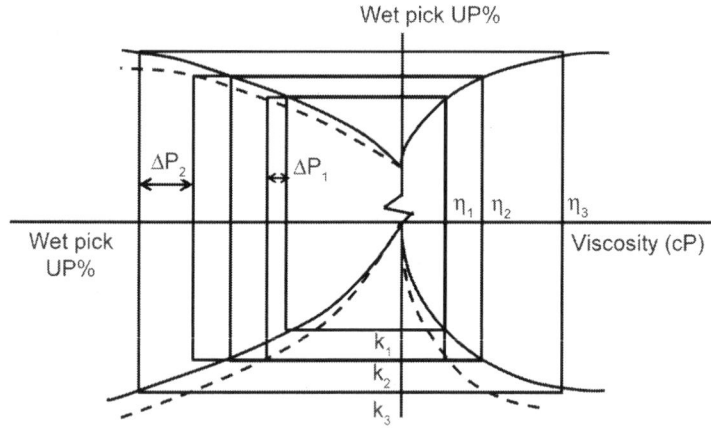

Fig. 4.52 Relation between viscosity, dry, and wet pick-up%

4.22.5   Concentration (%)

To explore the maximum advantages of high-pressure squeezing, a high concentration of size solution is required; but the high concentration in sizing paste is restricted by the viscosity factor because increase in viscosity would slow down the flow and therefore penetration of size into the yarn. The thin-boiling starches are better solution because they exhibit low viscosity at high concentration. Hence, it can be effectively employed in conjunction with high-pressure squeezing to reduce cost of drying.

4.22.6   The drying zone

Usually in conventional slasher sizing machines, drying of the wet yarn sheet is carried out by physical contact with a hot drum surface by conduction principle. To avoid any abrasion between drying surface and yarn sheet, both yarn sheet and hot surface must be moved at similar speed. Multiple-cylinder drying has replaced the double cylinder drying. The cylinders are driven at the desired angular velocity so that the surface speed equals the translational speed of the yarn sheet. As the various drying cylinders rotate on their axis, the yarn sheet in contact with the curved surfaces also moves along a sinuous curved path, coming in contact with a cylinder and leaving it after a wrap greater than π but less than 2π radians (in between π and 2π radians) before climbing on to the next cylinder. Next cylinder rotates in opposite direction to the previous one. In this way, this technique is employed to dry the two sides of the sheet alternately in multiple steps. Generally, the diameter and working width of drying cylinder is found to be 30–34" and with working width of 60–72" respectively. Super saturated steam is used to heat these drying cylinders and condensed steam is drained out in the form of water through appropriate tapes. In alternate arrangement, gas is also used for drying. It is evident that the most efficient heat transfer occurs when high quality (100%) steam at saturation temperature is condensed in the heat exchanger or process. The major extent of thermal energy in the steam, i.e. latent heat of vaporization is transferred when the steam condenses to water. Although, the steam is generally discharged from most steam boilers contains water molecules or mist that has not evaporated and this is called "wet steam". The wet steam has lower thermal transfer efficiency and not suitable for many commercial applications. To improve steam quality, wet steam can be superheated to create 100% quality or "dry steam" using a circulation heater.

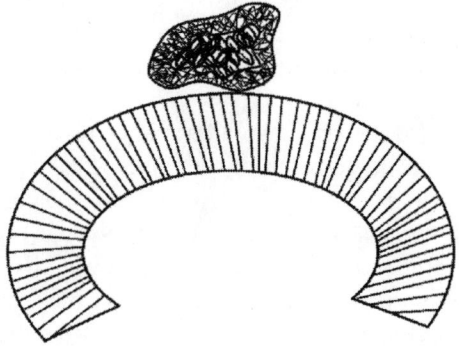

Fig. 4.53 Moisture gradient during drying

The direct contact between the size film-coated yarn sheet and the hot cylinder surface causes a localized instantaneous drying of the portion touching the cylinder. A moisture gradient develops in the yarn cross section (Fig. 4.53) that leads to a flow of the wet size across the yarn section and creates a migration action.

To prevent this migration phenomenon, it is desirable that moisture must be removed at equal rate over the entire section of the yarn. Otherwise one has to dry the two sides of the yarn alternately in many small steps, a strategy adopted in multi-cylinder drying systems. A typical layout of such a system is depicted in Fig. 4.54.

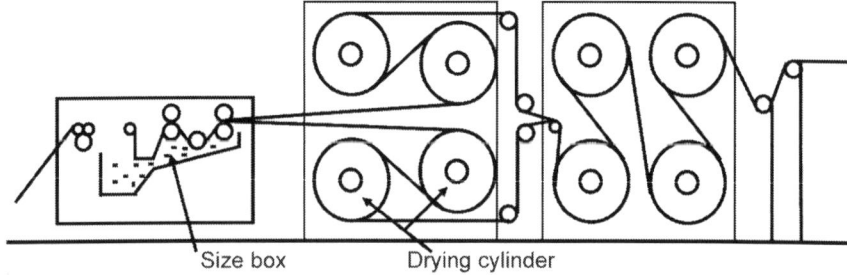

Size box   Drying cylinder

Fig. 4.54 Four cylinder double-section drying system

4.23   Various arrangements of multi-cylinder drying

In case of multi cylinder drying, various cylinders are arranged in different fashions and few of them are discussed here. The eight cylinders are in two blocks of four. The wet yarn sheet is initially split into two layers while entering the first block. This not only breaks the continuous size film encasing the warp sheet but sufficient space is also created between two adjacent yarns. After wet splitting, a continuous coat of size film develops around each yarn. If the yarns are fully dried in this state then the use of subsequent splitting zone would become meaningless.

Appropriate space between two adjacent yarns on a drying cylinder is likely to achieve a desirable result of the removal of steam from the drying yarns. Indeed an efficient removal of steam from the surface of the drying cylinders helps in two ways, first it lowers the vapour pressure to facilitate easy and efficient evaporation of water present in wet-sized yarns and second it prevents the formation of insulating steam layer called "steam pad" between the yarns and the cylinder. This steam pad lowers the thermal conductivity (indirectly drying efficiency) of hot cylinder surfaces. The wet splitting shares the load of each drying cylinder resulting in higher machine speed.

At the moment of detachment of the warp sheet from the first drying cylinder of drying cylinder series, the size film coat surrounding each yarn would still be wet and soft. In this condition, a part of size film does not peel off from the yarn and remain stuck on the hot cylinder during the process of detachment. The cylinders of the first drying block are coated with polytetrafloroethylene (PTFE). The very low coefficient of friction and high thermal stability of PTFE permits the process of drying in steps without sacrificing speed of sizing (sizing speed of 50–150 m/min). However, this additional layer of PTFE lowers the thermal conductivity of the cylinder (Thermal conductivity of PTFE at 273°K is in the range of 0.26–0.44, while the corresponding values for air are 0.025, for water 0.56, and for stainless steel 45–65 W/Km).

Indeed the thermal energy of steam injected into the cylinder is wasted considerably in overcoming the combined resistance of a layer of condensate on the inner wall of the cylinder, the wall of the cylinder itself and the layer of the PTFE as shown in Fig. 4.55. The thermal efficiency of the system is quantified and expressed by the ratio of the weight of water evaporated per unit time to the weight of the steam used. Generally, commercial practices utilize typically 50–60% thermal efficiency of steam. In case of multi-cylinder drying, the utilization of thermal energy depends on the extent of wrap of the yarn sheet around the individual cylinders that indirectly depends on the relative arrangement of the cylinders. Evidently, an offset arrangement of cylinders (Fig. 4.55) would yield higher wrap than a linear one (Fig. 4.54).

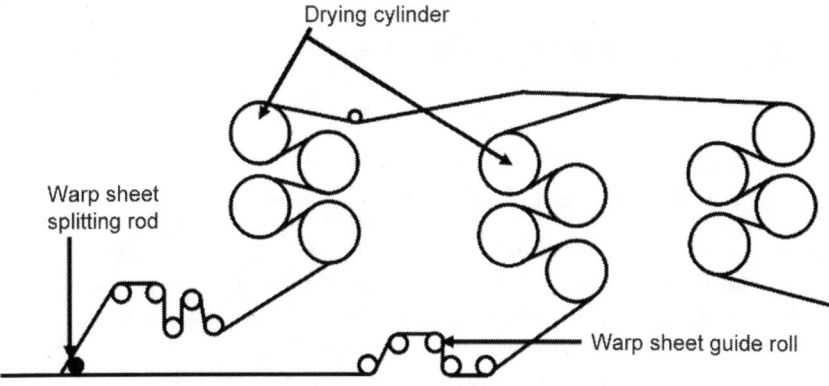

Fig. 4.55 Four cylinder triple-section drying system

The two layers of yarn as shown in Fig. 4.25 are dried to an extent by pre-drying arrangement before they are assembled again and dried together in the next block of four cylinders. The arrangement shown in Fig. 4.26 shows not only a different arrangement of cylinders but also two groups of four cylinders for pre-drying. Indeed, this arrangement also advocates about the use of two sow boxes for sizing the warp sheet in two parts separately. Such an arrangement yields better results for warp sheet of highly dense coarse warp threads.

Hot air drying and infrared drying have also been attempted in the past. However they have not proved to be commercially successful.

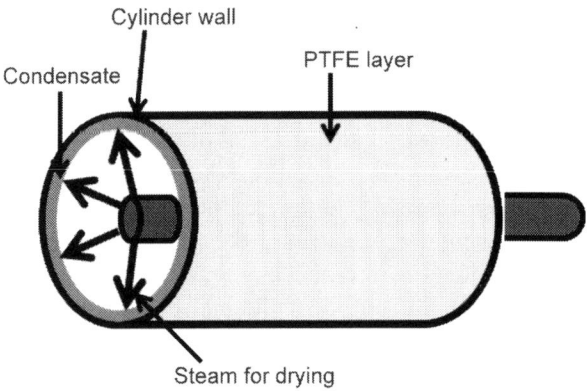

Fig. 4.56 Typical steam drying cylinder of sizing machine

4.24   Steam and condensate

The moisture of sized yarn sheet squeezing is removed by vaporization. This is done by steam heated cylinders in the drying section of the sizing machine. The heat transfer to the sized yarn sheet is made by direct contact to the outside wall of the cylinder.

Steam pressure measurement on drying cylinder of sizing machine
During the drying process, as the heat is extracted from the steam, a condensate film is caused on the inside wall of the cylinder. This condensate impairs the heat efficiency of the transfer to the yarn sheet. It is kept as thin as possible by skimming it off continuously and by measuring the pressure difference between the input and output. By using a VEGABAR 52, the effectiveness of the drying is monitored, and a siphon reduces the steam temperature from 160°C to below 120°C. The highly overload resistant CERTEC® measuring

cell ensures the necessary reliability even in case of pressure shocks in the cable. The cylinders can be installed in stacks of 6, 8, 10, or 12 cylinders and with two-three or more stacks to form a drying range. High speed sized warp sheet drying ranges are usually provided with an AC inverter-controlled drive with a flat belt or a chain and sprocket-drive to the individual cylinder. Efficient condensate removal from the cylinders through a well designed trapping system, coupled with dry saturated steam feeding, ensures excellent performance of the cylinder drying range. Cylinder drying range must be equipped with adequate size-rotary joints for steam feeding and condensate removal (from both sides of the wider-width ranges) and a specially designed trapping system with individual trap for efficient removal of condensate water.

Justified mechanical design, having potential to maximum cylinder surface utilization ensures lower energy consumption and higher efficiency. Following guidelines ensure maximum efficiency of drying range:
- Teflon coating on the cylinders
- Moisture indicator of controller acting on the steam inlet to the stack
- Curved-bar expander with guide roller or three curved-bar expander assembly before the first stack
- Big batching device or scray with big batching device for non-stop operation
- Steam feeding from both sides (mandatory for wider-width ranges)
- Steam line accessories for steam feeding and condensate removal
- Individual drive to each stack instead of common drive

4.25   Alternate drying arrangements

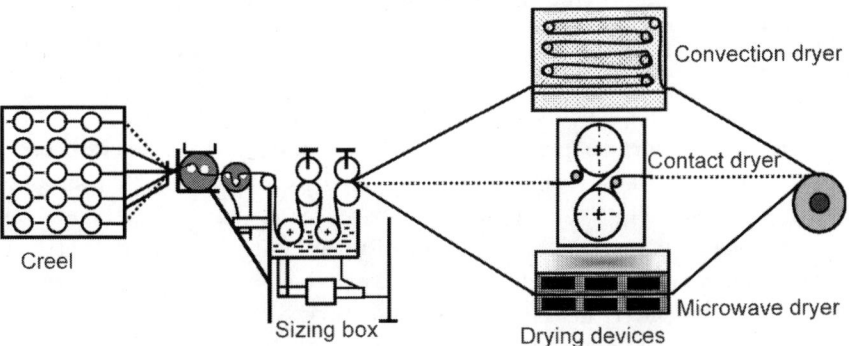

Sizing machine with three drying type of drying arrangements

Fig. 4.57 Alternate drying arrangements

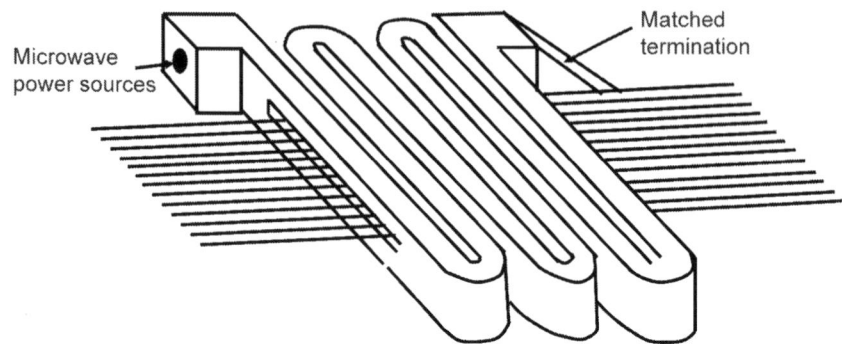

Fig. 4.58 Traveling wave applicator applied in the microwave device

4.25.1   Combined air/infrared drying for perfect drying

This system is used for perfect drying of sized yarns. The drying zone of this system is one metre long and contains a gas heater with a maximum output of 240 kW as shown in Fig. 4.57 and Fig. 4.58. This system has a drying performance of 40–50% at a processing speed of 60 m/min.

Low thermal efficiencies are common in many heating and drying systems. In many cases, infrared heating or hybrid systems that combine infrared and convection technology offer several advantages over conventional convection drying. Infrared radiation (IR) is commonly used to dry textiles and paper products, heat metals and plastics, and dry and cure paints.

4.25.2   Infrared (IR) versus convection heating

Medium-wavelength electric IR drying is especially well suited for the curing and drying of size films because these wavelengths correspond well with the absorption bands for water, which almost all size pastes contain. Additionally, since IR heating does not penetrate the surface very deeply and generally only heats the outer surface, it is applicable for drying coated products. Using IR heating can also facilitate the use of water-based sizing ingredients rather than solvent-based, which reduces emissions of volatile organic compounds. But perhaps the greatest advantage of IR drying over convection drying is that the IR emitters can deliver heat in exact amount directly to a specific point. IR is a line-off sight technology, meaning that it only delivers heat to the surface of an object that is in a direct line of sight from an IR emitter. Depending on the product, this can either be an advantage or a disadvantage. Flat parts of objects absorb energy best, while the complex curved parts require either more IR emitters or supplemental convection drying technology. Additionally, because IR drying only heats the surface of the material, it leads to significant energy savings when drying and curing coatings like size film on yarn surface, no heat is wasted on the underlying materials.

4.26   Hybrid convection/IR systems

Hybrid systems utilize both fuel-fired convection ovens and IR heating to create the most effective drying system for many textile applications. During drying of sized warp yarn sheet, when water-based sized yarns are dried, a layer of moist, saturated air forms near the surface of the product that produces the "boundary effect". This air layer tends to hinder the drying process. In one configuration of a hybrid system, the movement of air in the convection drying system in the initial drying stage helps to force this moist air layer away from the warp sheet while allowing the IR heaters to finish the drying in a second drying stage. This combination of convection and IR technologies works well for drying sensitive materials where the more controlled IR drying in the last stage helps to avoid damage to the material from excessive heat as shown in Fig. 4.59.

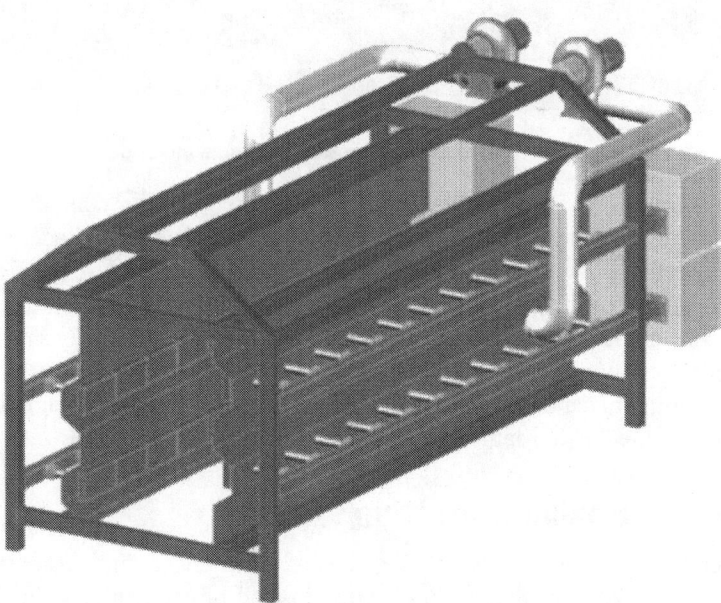

Fig. 4.59 Combined air/infrared drying

4.27   Residual dryer

This type of drying system consists of a cylinder drying which assures smooth yarn surfaces and uniform yarn-tension setting in both wet and dry zones. An individual motor-drive arrangement at each cylinder prevents the overstretching of yarns as shown in Fig. 4.60.

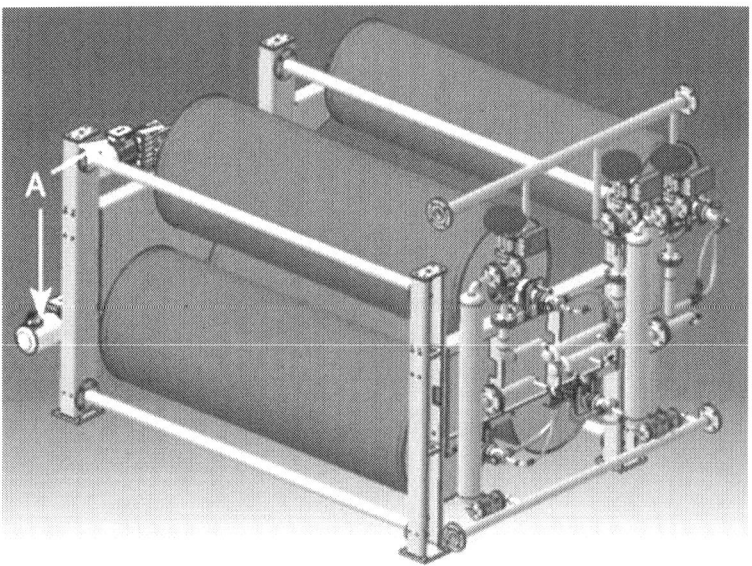

Fig. 4.60 The residual dryer with individual motor drive

To optimise the process control in weaving preparatory, "compact size" is very useful machine which has the facility to re-circulate the cylinder stream. This system has facility to increase heat transfer and a higher drying efficiency. Recirculation of the heat-transfer medium guarantees uniform temperatures, constant temperature control, and the smallest amount of condensate in the cylinders as shown in Fig. 4.60. After introducing the "compact size", short warp sizing machine, KARL MAYER is also promising the continuous success of the Gir-O-Matic sample warping machine.

4.28   Flexibility in sizing system

Basically, there are two ways of sizing of warp sheet: first, conventional hot-sizing system by spray impregnation instead of conventional dipping of warp sheet in size paste; and second, the cold sizing technique. When sizing process is performed by tangential sizing technique without using heat, the sizing ingredients are applied by means of double-kiss rollers. A separate unit with its own circulation system and different heat-supply concept is discussed for each of the two above-mentioned application systems. Whereas hot sizing operates with direct and indirect heating, cold sizing only uses indirect heating.

4.29   The splitting zone

The yarns in warp sheet after application of size paste coming out through drying chamber may stick with each other. The degree of sticking depends on the efficiency of pre-dryer assembly. The function of pre-dryer is to prevent the entire warp sheet from being covered in a continuous uniform coat of size film. As the yarns would need to be separated out for the subsequent operation, safe splitting of warp yarn is essential. The yarns in the warp sheet coming out of the drying section adhere to each other to a degree depending on the efficiency of the pre-dryer. In the absence of a pre-dryer, the entire sheet would be encased in a continuous coat of sizing material. As the yarns would need to be individualized for the subsequent operation, the coat of size film needs to be broken at many places. The lease bands introduced during warping are introduced to split the sized sheet in steps. Rods are inserted in place of the lease bands as depicted in Fig. 4.61 and held in position by suitable brackets.

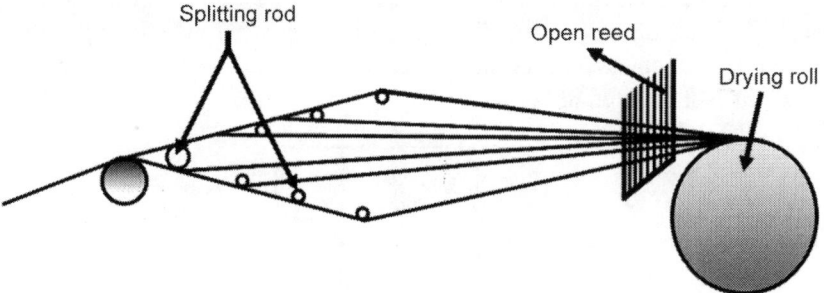

Fig. 4.61 Splitting of yarn sheet (side view)

During splitting, some amount of size material would also flake off as waste. Simultaneously, a large number of relatively long protruding fibers on surface of yarns and the fibers bridging the neighboring yarns would also get broken into smaller pieces. In this way, splitting becomes beneficial in so far as reduction of hairiness is concerned. However, this process may exhibit an uneven and possibly even a ragged and damaged size film around each yarn. This is illustrated in Fig. 4.62

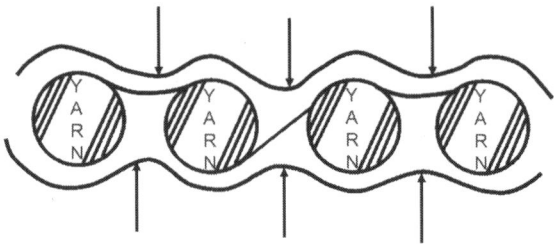

Fig. 4.62 Splitting of yarn sheet

4.30   The beaming zone

The individualized sized yarns after splitting in warp sheet form would finally be wound on weaver's beams of desired width. Generally, one sizing machine feeds to a large number of looms of varying width. The headstock of a sizing machine must have the flexibility to adjust according to the widest as well as the narrowest weavers beam. The beam is driven at one end by a spindle, to which the beam pipe is fastened by clamping the same in a three jaw chuck and the other end of the beam pipe is provided with a support to permit free rolling. This supporting end has facility to move in and out to accommodate beams of various widths as also to permit doffing of the full beam. The width of the warp sheet must match to the width of beam on which it is to be wrapped. Hence, the sized warp sheet is at first passed through an expanding reed/comb, by means of which the thread spacing is adjusted to the desired value. This is illustrated in Fig. 4.63.

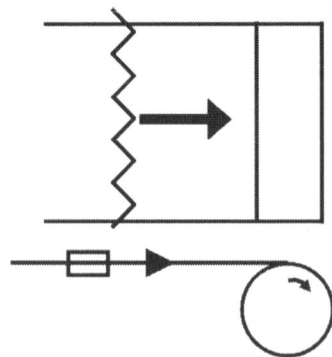

Fig. 4.63 Adjustment of width of warp sheet

To maintain constant winding speed of weavers beam, the angular speed of the spindle-driven beam needs to be reduced continuously with increase in beam diameter. Continuous monitoring the yarn tension and controlling the transmission ratio between the driving motor and the spindle by means of a positively infinitely variable (PIV) box are used to achieve constant winding speed. The hardness of the beam is controlled by maintaining the yarn tension at a desired level and by applying a press roll on the beam being wound. A soft beam is not suitable for weaving because the parallel wound yarns may collapse under fluctuating weaving tension. Moreover, a soft beam would contain less amount of material thereby causing frequent interruptions to the subsequent production processes. On the other hand, a very hard beam may raise the strain level in yarn layers close to the beam pipe beyond a critical level, thereby causing permanent deformation and changes in yarn property and can also cause rise in warp yarn breakages during weaving process due to loss of resiliency of the warp beam. Hence, a precise control of yarn tension and also pressure on beam throughout the beaming process is crucial to form a quality weaver's beam.

After passing through the expandable comb, the yarn sheet is sensed by a moisture meter to measure the value of moisture content against a reference value. Depending on the type of error, the speed of the entire machine is either raised (moisture level below the lower limit) or lowered (moisture level above the upper limit) to adjust the residual moisture content in weavers beam. A modern sizing machine is equipped to produce beams up to a width of 6000 mm and diameter of 1600 mm. The power required by such a machine can be around 30 kW. Presently, modern sizing machines are able to produce a beam up to a width of 6000 mm and diameter 1600 mm. These machines require 30 kW power to run successfully.

4.31   Controls on modern sizing machine

A modern sizing machine is equipped with the following types of control systems:

4.31.1   Stretch control

During the process of sizing, yarns are subjected to stretch between the creel and size box, the size box and drying cylinders, the drying cylinders and draw rolls, and the draw rolls and final weavers beam. Depending on the type of yarn being sized, the net stretch has to be kept under control so that the yarn does not lose elasticity while being sized. For example, a cotton warp sheet is expected to undergo a net stretch of around 1–1.5%. A precise control on the speed of the driven elements is a prerequisite for such a precise control. The stretch meter can be used to indicate the % stretch between various zones.

4.31.2   Tension control

The tension regulation in running yarns is based on following facts:
- Tension meters are employed to measure tension in warp yarns. The yarn stretch is affected by the yarn tension.
- Tension control is directly used to regulate the angular velocity of beam. Moreover, if the yarn property is known exactly then stretch can be kept under control by controlling the tension.

4.31.3   Moisture control

Moisture meter reading just prior to beaming is employed to control the overall speed of sizing without disturbing the speed ratios of critical driven elements because by controlling these parameters, residual moisture content can be adjusted as per the requirement of the process.

4.31.4   Pressure control

The pressure in the nip of squeezing rollers affects the wet pick-up and hence the cost of drying. A better anchorage of size film in the yarn body reduces size flake-off, thereby ensuring better protection. On the other hand, too high a pressure may flatten the yarn affecting its appearance and subsequent behavior. The amount of pressure to be applied depends also on the material being sized, density of the sheet, and total required size add-on %.

4.31.5   Temperature control

The effective temperature control at various zones of sizing machines not only improves the productivity, but also the homogeneity of size paste coating on yarn surface. Following facts reveal the importance of precise temperature control in sizing process:
- The temperature of the PTFE coated cylinders is lower than those of the succeeding layers when filament yarns are sized; while no such difference is maintained during sizing spun yarns.
- Cotton yarns are dried at higher temperature than polyester cotton blended yarn.
- Indeed, the temperature of each cylinder needs to be controlled and maintained at different levels for different materials.
- This depends on the amount of water to be evaporated by each cylinder and the undesirability of over drying a yarn surface and causing excessive migration during the initial stages of drying.
- Indeed the migration of size in a filament yarn, that lies flatter on

the cylinder in wet state than a spun yarn, might even be across the section instead of being along the periphery.

4.31.6   Level control

This system ensures a constant depth of immersion of the yarn sheet within the size paste and contributes to uniformity in wet pick-up along with other size box parameters.

4.32   Special sizing systems

4.32.1   Foam sizing

Yarn sizing is the costliest process of weaving preparatory due to high requirement of energy and raw materials. The major energy requirements in the sizing section are for heating the water during size cooking and evaporating the water during yarn drying. Assume for a moment that water is replaced by air then there will be significant reduction in energy consumption. This fact was utilizes in the development of "foam sizing" (Walker et al. 1980; Schutz et al. 1989; Wamboodri 1986; Perkins et al. 1982a; Perkins et al. 1984b; Rajendran et al. 1984).

Principle of foam sizing
Foam is a mass of tiny bubbles separated by thin liquid films. By introducing air into a concentrated size solution, a given volume of liquid spreads over a large surface area. Very small amount of highly concentrated solution can spread onto a large area of textile material because of the high ratio of liquid surface area to liquid volume. In conventional sizing, water is the medium by which size particles are held in solution or dispersion form and water also carries the size particles to the textile material. In foam sizing, water is not completely eliminated and still used to hold the size particles, but foam (combination of air and water) is the delivery medium in place of 100% water as used in conventional sizing process.

Foam sizing system
Generally two types of systems are being used to apply foamed chemical solutions to textile materials. The first uses a horizontal pad, a knife over-roll coater or other appropriate applicator to apply relatively stable foams to the textile material. The applicator collapses the foam bubbles after the foam has been uniformly applied to the yarn surface as shown in Fig. 4.64. Chemical add-on is determined by blow ratio of foam and the volume of foam on to the yarn. Both parameters are critical and must be controlled precisely. The

second type of foam applicator uses relatively unstable foams that collapse at a predetermined rate. In this system, the required quantity of foamed chemical solutions meters on to the material. Unlike the first system, the foam normally is not collapsed by squeezing or other mechanical means.

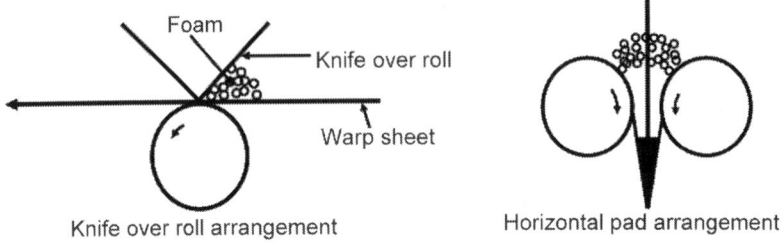

Fig. 4.64 Foam size application by knife

Examples of foam sizes
The conversion of non-starch size material like carboxy methyl cellulose (CMC) or PVA into foam is comparatively easy (Rajendran 1984). PVA with 25% solid content foams easily. Some researchers studied foam formation possibilities of fully hydrolyzed and partially acetylated grade PVA. PVA with low viscosity has a greater propensity to produce low density (high blow ratio) foams than higher viscosity size materials. Liquid polyester size, foams easily to produce a blow ratio greater than 10:1 with hand mixer. The foaming agent is usually added just prior to foaming. Foaming chemicals should not react with the material to be sized and it should have a good dispersing property in the cold water.

Merits of foam sizing (Perkins a and b, 1982, 1984)
1. Energy saving in drying of sized yarn and enhanced productivity
2. Size cooking kettle and size box are not required
3. This process is comparatively very economic from energy saving point of view from conventional yarn sizing processes
4. Due to low drying requirement, less specious drying chamber saves the working space
5. Percentage add-on reduces with improved size utilization
6. Saving of size material
7. Easy desizing
8. Less bridging and yarn hairiness
9. Less shedding or cleaner loom shed and higher weaving efficiency
10. Due to less heat utilization, the working conditions in the size room will be less exhaustive

Table 4.1 Comparison between conventional sizing and foam sizing.

Parameter/ particular	Conventional sizing	Foam sizing
Size add-on % desired	12.0	12.0
Wet pick-up obtainable %	100.0	40.00
Size % in sizing paste %	12.0	30.00
Water to be evaporated kg/kg of yarn	0.88	0.25

Foam generation

Foam generation is the result of mixing of gas-phase and liquid-phase in a foam mixer under influence of appropriate kinetic energy. For foam sizing point of view, the liquid phase is usually water containing surface-active agents and chemicals to be applied to the yarn surface. In most cases, air is used to form gas phase. The structure of foam may be characterized by the blow ratio and the bubble size distribution. The blow ratio is the ratio between volumes of gas to the volume of liquid in the foam. The degree of dispersion of the foam is measured by the bubble size distribution. The gas and liquid flow rates supplied to the mixer regulates the degree of dispersion of the foam, while energy supply to the mixer controls the bubble size. A high energy supply per unit volume gives rise to small bubbles in the mixer. The two types of foam mixers used to generate the foam for foam sizing are rotor-stator mixers, and static mixers. The energy supply to rotor stator mixers is calculated by the energy transferred from the revolving blades to the foam mixture while in static mixers, the energy supply is related to the flow rate of the foam. The energy consumption in static mixers depends to a great extent on the production rate of foam (which is proportional to the foam flow rate in the mixer). The rotor-stator mixers are capable of producing foam with a consistent structure over a wide range of foam production rates as discussed by Kroezen et al. (1986).

Studies on foam sizing

Various merits of foam sizing attracted researchers to investigate it precisely for process optimization. Perkins et al. (1982) studied the characteristics of different size materials including polyvinyl alcohol, carboxymethylcellulose, derivatized starches, acrylic polymers, and polyesters to explore the influence of size formulation viscosity and concentration on foam ability. Perkins concluded that the properties of foam sized polyester/cotton yarns are equivalent or superior to those of similar yarns sized conventionally.

Namboodri (1986) utilized the Reed-Chatwood applicator (a horizontal padder) to size warp yarns by foam sizing to collapse the foam at the nip of the padder to maintain size uniformity with lower wet pickup. This study also revealed that the size add-on is controlled by the blow ratio and the solid content in size foam. Foam sizing exhibited bridging of yarns and reduced yam hairiness at low wet pickup.

4.32.2   Dye sizing

The use of colored warp thread has registered a massive growth in world textile industry to weave some typical fabrics like denim. Indigo dye is used to dye the warp threads. Warp threads may supply from indigo dyed cones. The yarn dyeing may take place on cone by winding the yarn on perforated conical shell under low tension and pressure so as to create soft package suitable for dyeing. These packages can be put in a closed chamber under pressure in which the dye liquor is circulated. However, such a dyed cone needs to be rewound for warping and sizing.

To outwit such a tortuous process, sizing machine manufacturers have been designing machines to dye the warp sheet and then size the same at one integrated process. It is clearly evident from this process that only limited type of materials can be handled on this system.

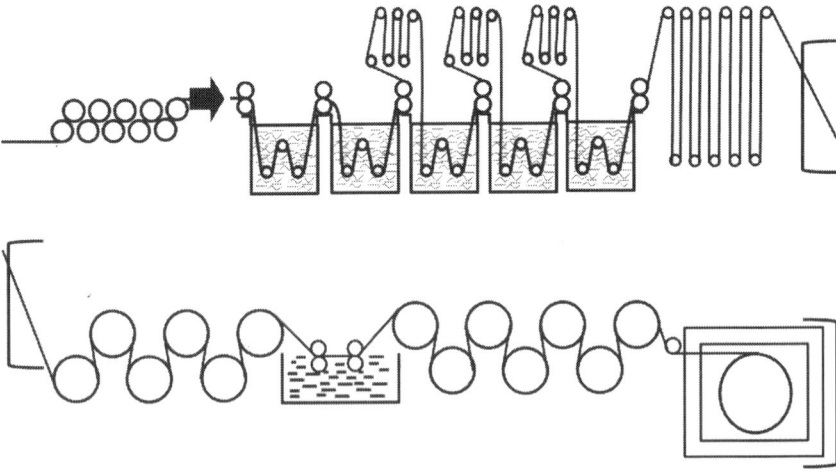

Fig. 4.65 Dye-sizing machine for deni

Dyeing cotton yarns by reactive dyes need curing at elevated temperature (ca. 130°C). The curing takes place with the help of hot air or steam and the curing is a relatively slow process. On the other hand, indigo dyeing requires oxidation at room temperature and does not need a long dwell time for curing and other processes. Hence a dye-sizing machine for denims would need many vessels and guide rolls, while dye-sizing cotton or polyester cotton yarns with reactive and disperse dyes would need additional provision of curing chambers, occupying a considerable space for ensuring the required dwell time in series as illustrated in Fig. 4.65.

4.32.3   Single-end sizing

As per Tsudakoma and Benninger group, the single-end slashing machines are commercially available currently. However, these machines are primarily used for sizing filament yarns.

The single-end sizing method is a special method for some typical applications. Many of the conventional sizing techniques are not appropriate for short-run production and archaic and slow by present production standards. For example, the conventional slasher sizing process is based on 150-year-old technology with maximum production speed of about 150 m/min. Very huge set-up, high energy wastage, and slow run make this process uneconomic. The conventional sizing process is a self-metered coating process. All these conventional sizing processes apply size to warp sheets by moving the warp strands from a series of beam creels through a size box containing an aqueous solution of size, which coats the warp yarn. The yarns are placed closely on the drying cylinders, causing the sized yarns to stick together forming a solid sheet as they leave the drying cylinders. The separation of stick yarns is carried out by passing them over and under lease rods. This separation of yarns leaves pulling of fibers out from yarn surfaces as bridged size films are torn apart by the bust rods, producing a hairy, non-round yarn as described by Ok et al. (2007) and Lord (1982).

Various efforts to modify the sizing process and sizing materials are continuously made. Trauter 1 and 2 (1984 and 1996) explained that the wetting with hot water before sizing, modifying the fiber surface and modifying the size formulation improve the effectiveness of the sizing process. Conceptually, single-end sizing is an effort to discard the problems related with the conventional sizing to individual yarns and drying each yarn separately. Benninger and Tsudakoma offered single-end1 sizing machines primarily for sizing filament yarns. In single-end sizing, the method of applying size is similar to conventional sizing; in that, size is applied by pulling individual yarns through a size box and squeezing them to remove excess size. The method of applying size is similar to conventional sizing, in that size is applied by pulling individual yarns through a size bath and squeezing them to remove excess moisture. However, the hot-air dryer is used for yarn drying in Tsudakoma's and Benninger's single-end sizing machines.

Ok et al. (2007) has designed a single-end sizing process referred as slot applicator to apply size to individual yarns in a metered fashion to give uniform application of size in order to improve the flexibility and productivity of sizing process. For size application, the yarn is passed through a long slot and multiple slots are used to process multiple ends of a warp sheet. The viscosity of the size solution had to be kept sufficiently low so that good size coverage was achieved.

The major advantages of single-end sizing process are as follows:

- The yarn sized by this process is much less hairy than yarn sized by the traditional sizing method, and yarns of reduced hairiness are preferred particularly for air-jet weaving.
- The maximum speed of this process is 50–360 m/min depending on the type of yarn.
- Since the size solution is applied in a metered fashion, the system allows quick style changes with easy adjustment of size add-on for different styles.
- Single-end sizing reduces the amount of water applied to yarn; therefore, it reduces energy requirements during drying.
- Polyacrylamide size is recyclable, which is used on this process; the process is environmentally sustainable.
- The elongations of single-end sized yarn were lower than those of un-sized and commercially sized yarn.
- The tenacities of single-end sized yarn were similar to those of commercially sized yarn.
- The microscopy study also showed that the single-end sizing process produced smoother, less hairy yarn than the commercial sizing process.

This process basically involves a high-density package creel from which yarns are withdrawn and formed into a sheet, which is then sized straightaway. As a creel can hold a very limited number of yarn packages, the resultant thread density in the sheet is very low such that the distance between adjacent yarns in the sheet is a multiple of yarn diameter. Hence the resultant size film forms securely around each yarn, doing away with the necessity of splitting the dried sheet. Consequently single-end sizing is highly beneficial for filament sizing process, as individual filament may rupture during splitting process. However the beam coming out of such a sizing machine would contain only a limited number of threads, insufficient for a woven fabric. Hence, a separate beaming machine subsequently assembles required number of such sized beams into the final weavers beam.

Such a process is illustrated in Fig. 4.66.

In a modified version of this process, yarns from the creel are warped to a beam; the beams are individually sized and assembled by a separate beaming process. This modification takes care of interruptions in the sizing process caused by filament/yarn breakages as a result of yarn unwinding from creel. This modified process is easy to implement with conventional machine set-up.

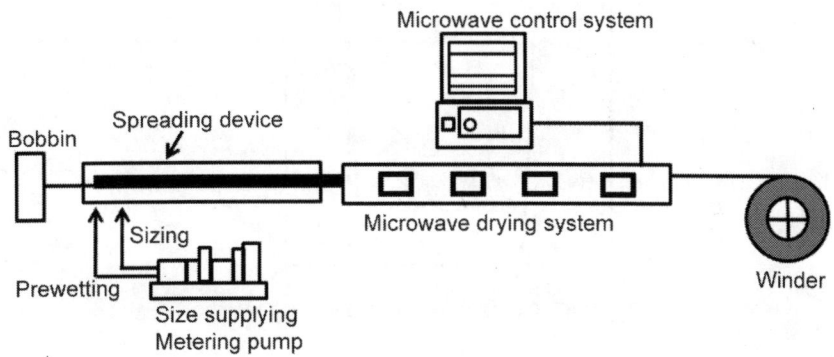

Fig. 4.66 Single-end sizing process

CCITech single-end sizing machine

CCITech Taiwan has realized the problem of researchers and developed a single-end sizing machine for small sample manufacturing. This machine is able to size any minimum quantity of yarn. This machine does not have size cooking facility. Hence, separately prepared size paste is filled in the tiny size box of the machine. The supply package is placed on bottom side of the machine. The yarn is carried up to size box with the help of thread guide. Inside the size box, a clamp with roller works as immersion roller to dip the yarn in size paste properly. With the help of thread guide, the wet yarn reaches to a pair of positively driven squeezing roller (nickel-coated stainless-steel roller at bottom and rubber-coated roller at top) as shown in Fig. 4.67. After squeezing, the sized yarn enters in hot-air drying chamber which is equipped with epicycle gearing to push forward the dried yarn continuously. The yarn separation from drying swift takes place by the disk-type self-compensating tensioner, and finally the sized yarn winds on the package with the assistance of grooved roller-driven winding system. The rotary traverse is provided by the rotation of winding drum. Sizing speed and hot-air temperature is controlled digitally. To maintain the viscosity of size paste, a continuous stirrer and hot plate is attached with size box.

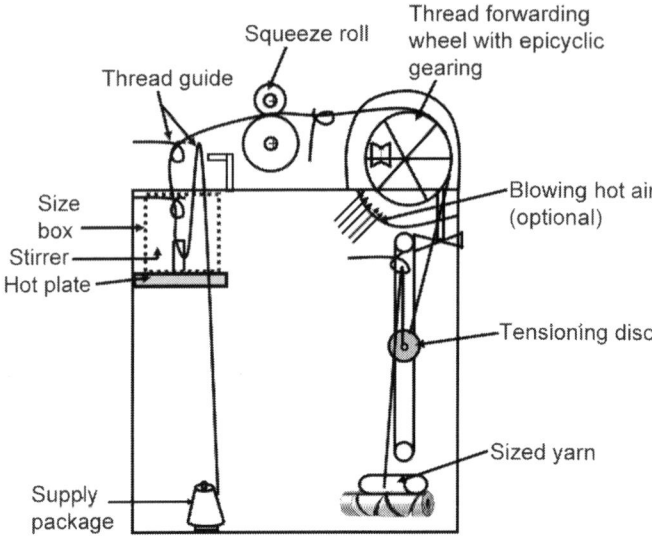

Fig. 4.67 CCITech single-end sizing machine

4.32.4   Hot-melt sizing

The concept of hot-melt sizing is entirely different from traditional wet sizing process. The hot-melt sizing process is integrated with the warping process by the application of a molten, 100% active, size material to the warp yarns as shown in Fig. 4.68. The speed of hot-melt sizing machine remains around the average warping speed of conventional machines.

The concept of hot-melt sizing is illustrated in Fig. 4.68, which is located between the warper creel and the warper. The warp yarns drawn from a warper creel pass over the top of an internally heated rotating drum. Deep and closely spaced parallel grooves are formed on the surface of this roller, each groove being meant for one yarn strand, allowing good surface contact with the yarn. The size is applied by pressing a block of solid size against the grooves of the hot cylinder.

The molten size material is transferred on to the warp threads when they are in contact with the grooves of the roller. The groves separate the ends during size application and also allow a good surface contact with the yarn. The grooved surface rotates at a speed of about 10 rpm in the direction of warp sheet movement from creel to warper. The grooved roller is uniformly heated to ensure a temperature of about 150°C. The warp threads moving in a sheet form will have a higher surface speed, compared to the groove roller surface speed. This difference in surface speed is used to wipe the size material from grooves to the threads and reduces the yarn hairiness. As the warp threads come out from groove area, they immediately come in the contact of surrounding cool air. Consequently, the size sets very quickly on yarn surfaces to bond the protruding fibers on yarn surfaces.

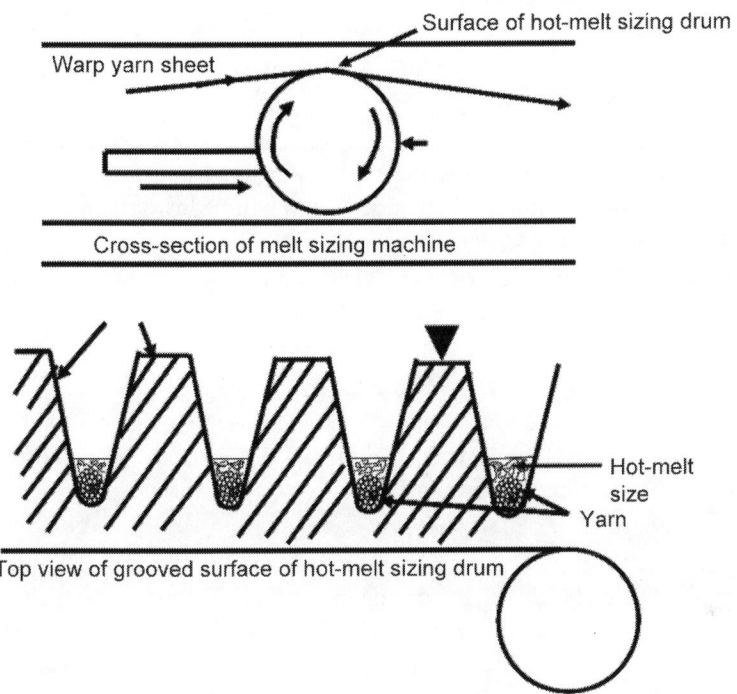

Fig. 4.68 Hot-melt sizing

The high temperature of grooved surface does not affect the thread properties, because the yarn is in contact with the heated grooved roller only for a short time span (0.0072 s at about 600 m/min warping speed). The percentage add-on is governed by hot-melt sizing roller speed, time of contact of yarn with groove, warping speed, groove roller temperature, and the feed rate of size to the grooved roller. Groove depth remains three to five times the diameter of the yarn to get optimum size wipe.

The melt blend of ethylene/vinyl acetate copolymer with microcrystalline paraffin, or hydrogenated tallow wax is a typical example of hot melt size ingredient. Alternately, a composition made of an intimate mixture of a water-soluble film-forming thermoplastic polymer and a melt-miscible, solid modifier selected from the class consisting of aliphatic and aromatic carboxylic acids and nonpolymeric polyhydric alcohols or phenolic acids/ polyhydric phenols, wherein the proportion of polymer to modifier is about 90:10 to about 50:50 on a weight basis and can also be used as the said size material. The size can be removed from the yarns by aqueous or alkali desizing.

The advantages of hot melt size are as follows:

- *Low energy consumption*
  Since there is no water to be removed from sized yarn, the energy required to size by this method is about 80% less than is needed for aqueous sizing.
- *Low investment cost*
  The investment cost for a hot melt sizing unit and for a machine, required for assembling the warper beam will be lower than those for a conventional sizing and the size cooker.
- *Elimination of cooking area and its dumping*
  There is no size solution to prepare and dump. A hot melt size is available in ready to use form from the supplier.
- *Superior quality sizing of warp yarn*
  Due to the separate-end sizing of each yarn, sizing quality is superior to that achieved by conventional sizing. Also, the method provides better fiber laying on yarn surface.
- *Greater speed of size application*
  Since there is no water to be removed, the speed at which size is applied is not limited by drying potential. The rate at which the molten size sets up to a non-tacky state determines the sizing speed, the set-up rate of one second or less are achievable without any assistance of any type of auxiliary equipment, allowing application speeds equivalent to conventional warping speeds.
- *Resistance to shedding*
  It is a result of the excellent elastic properties of the advanced size materials.

Requirements of hot-melt size
It is remarkable here that the hot-melt size is not simply a waxing process, and it must have the following characteristics:

- Melt temperature remains about 260–310°F
- Hot-melt size must have ample resistance to heat degradation
- Hot-melt size should not block the groove
- Size should be desized by conventional (aqueous desizing process) desizing process.
- Process should allow rapid set-up

- Size must be matchable with fibrous material to improve its tensile strength and elongation
- Material should be sized at acceptable cost

Disadvantages of hot-melt sizing process

- *Size recycling*

  The projected cost of aqueous hot-melt size recovery does not justify size recycling.
- *Operator training required*

  The hot-melt sizing and size material are unconventional, operator training is requisite.

  However, the method is less complicated and easier to operate than conventional sizing method.
- *New equipment requirement*

  Investment in new equipment is required to change over from conventional to hot-melt sizing.
- *Different size*

  The size material is not a wax and is unlike any conventional aqueous sizing materials.

4.32.5   Semi-dry sizing

The semi-dry sizing is also an effort to reduce the energy consumption in drying of wet-sized yarn in conventional sizing process as shown in Fig. 4.69. The warp beams are mounted on creel as per the requirement of weavers beam. The yarn sheet passes by the electrical dropper that works as thread-stop motion and then sheet is divided by both vertical and horizontal dividing rods for better separation of warp yarns. Further, the yarn sheet passes over sizing roller that is partially dipped in size paste. This arrangement applies a minimum amount of size paste on warp threads. The press roller squeezes the extra amount of size paste and distributes the size paste on yarn. Cooling pipes are used for wet splitting of sized sheet. Ahead, the warp sheet enters in drying chamber for drying. The drying chamber is divided into various sections to control the drying temperature more precisely. The warp threads again pass through electric dropper that works again as thread-stop motion. Again, lease rods are inserted to separate the sized yarns from each other. Ahead, the sized yarns pass through yarn sheet arranging reed to adjust the width of sheet as per space on weavers beam. Finally, the sized yarn sheet is wound on weavers beam. The major advantage of this system is the reduction in energy consumption utilized for drying of wet yarns, and higher productivity in comparison of conventional slasher sizing processes.

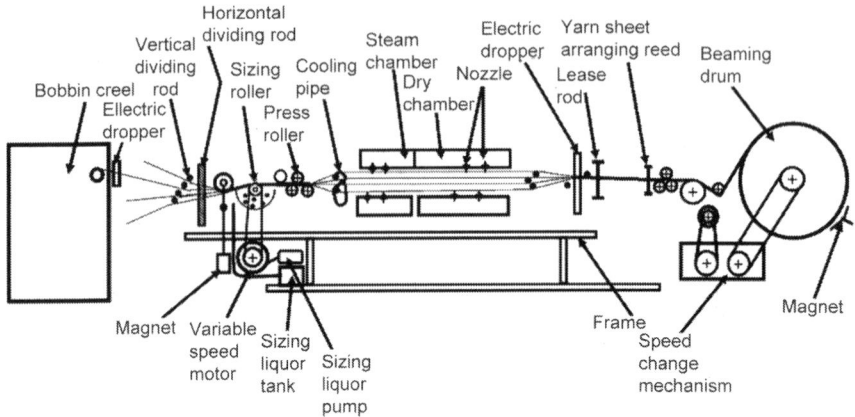

Fig. 4.69 Semi-dry sizing

4.32.6 Solvent sizing

The solvent sizing machine is shown in Fig. 4.70. The increasing energy cost and pollution control legislation have resulted in the development of alternate sizing systems and solvent sizing is one of them. In this system, chlorinated hydrocarbon is used as solvent in place of water. Polystyrene resin is used as sizing material and the solvent is chlorinated hydrocarbons like 111 trichloroethane, perchloroethelene, and trichloroethylene, etc. Since the chlorinated hydrocarbon has very low latent heat of vaporization, the energy saving will be enormous. In case of solvent/solvent process, solvent dispersible sizes are removed by a solvent desizing system in processing plant. The size is being recycled and returned for reuse.

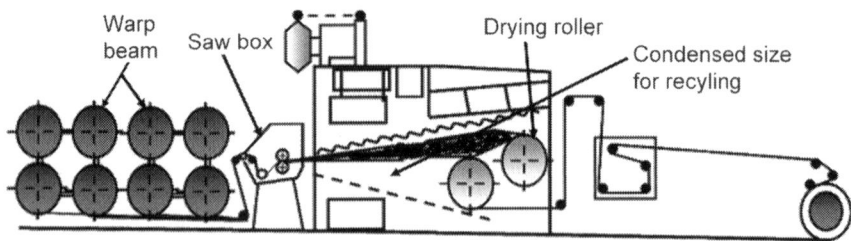

Fig. 4.70 Solvent sizing machine

The warp sheet is passed through the size box where size material is dissolved in the solvent at room temperature. The warp sheet coming out of the size box is wet split before being taken to the cylinder for drying .The wet warp sheet comes in contact with the heated solvent vapor which partially dries them. The whole unit is completely enclosed. The two condenser coils condense the solvent vapor and take it to storage tank for further use. The size is mixed in a simple tank at room temperature using a high-speed propeller-type mixer.

Bowman et al. (1996 and 1998) developed a sizing process using liquid carbon-dioxide (CO_2). The major advantage of using liquid CO_2 as a solvent is that it has the potential of reducing the energy requirement during slashing but none of the existing size formulations, such as starch or polyvinyl alcohol (PVA), is soluble in the liquid carbon-dioxide. This process required new size ingredients to be used. This method requires a technically appropriate method to handle liquid CO_2 as stated by Bowman et al. (1996 and 1998).

However, none of the existing size formulations, such as starch or polyvinyl alcohol (PVA), is soluble in the liquid carbon-dioxide; therefore, new size formulation must be developed for the method to be used. This method will also require the handling of liquid CO_2, not typically done in textile processing.

Essential requirements of solvents for solvent sizing
- Solvent must be safe and easier in handling
- The latent heat of evaporation and boiling point must be low
- Solvent should not affect the properties of the textile fiber or sizing material
- The solvent should be environmentally acceptable and easily recoverable

Advantages of solvent sizing
- Reduction in energy cost: Since the energy requirement for evaporation of solvents (chlorinated hydrocarbon) is low in comparison of water, the energy required to dry a warp sheet is also very low. Besides the obvious saving in production cost, this also results in higher production from a compact unit showing saving in space and capital cost (Tauter 1980; Platt sizing 1980; Annon 1977a; Anon 1977b).
- Versatile in terms of multi-product weaving: The solvent sizing is very versatile, same machine can be used for sizing a wide variety of yarns made out of synthetic and natural fibers, which is due to the capability of solvents to wet all kind of fibers easily. It gives better sizing because the hydrophobic fiber like polyester gets instant wetting.
- The heat required to evaporate the solvent is a fraction and usually less than 1/8th of that water.
- Suitable with fibers of low wet strength: Regenerated fibers have lower wet strength in aqueous medium compared to their dry strength. But when

viscose or viscose-blended yarns are impregnated with solvents, the loss in strength is zero.

- The level of pollution is zero because the solvent is evaporated, condensed, and recycled in a closed system. There is zero discharge of wash-down liquor as drainage which saves material cost. Generally, 80% of the size can be recycled.
- This process is especially suitable for viscose fiber which loses its strength in aqueous medium.
- No detectable size migration: In solvent sizing no detectable size migration onto the drying cylinder occurs, which is due to the low surface tension of the solvents. Solvent size coupled with pre-dryer, eliminates any possibility of size migration.
- The solvent sizing system is incorporated with wet-splitting and partial-drying. So that final warp sheet will be less hairy.
- Abrasion resistance of the solvent sized yarn is higher than that of its counterpart conventionally sized yarn.
- High production rate: High production rate is possible due to the quick drying of the warp sheets.

Disadvantages of this process are as follows:
- Toxicity of chlorinated solvents: The chlorinated solvents are highly toxic, causing environment related problems (pollution).
- Higher investment cost: Due to special safety regulations, the investment cost is high. Also additional energy is required for the vacuum system.

4.32.7   Cold sizing

Cold sizing process is specially designed for heat sensitive fibers like wool because during drying of wet sized yarn, fiber goes for severe heat treatment and results in deterioration in the basic characteristics of the fiber. The concept of cold sizing is given in Fig. 4.71.

Cold sizing agents are blends of polymers based on PVA with high adhesive power and anti-static agents dissolved in water. These are available in different viscosities, concentrations, and adhesive power. Application of cold sizing agent is carried out with a waxing system located between the warping drum and beaming machine of a sectional warping machine. The size is applied indirectly by two-sided tangential applicator. Through rotation of the rolls and simultaneous application of pressure, a finely adjustable size film is realized on the warp. By adjusting the difference in speed between warp sheet and applicator roll, the pick-up can be adjusted. Infrared radiators dry the warp sheet without contact. Sometimes skying (by air contact only) is also used for drying at very slow speed. The advantages of this drying method are

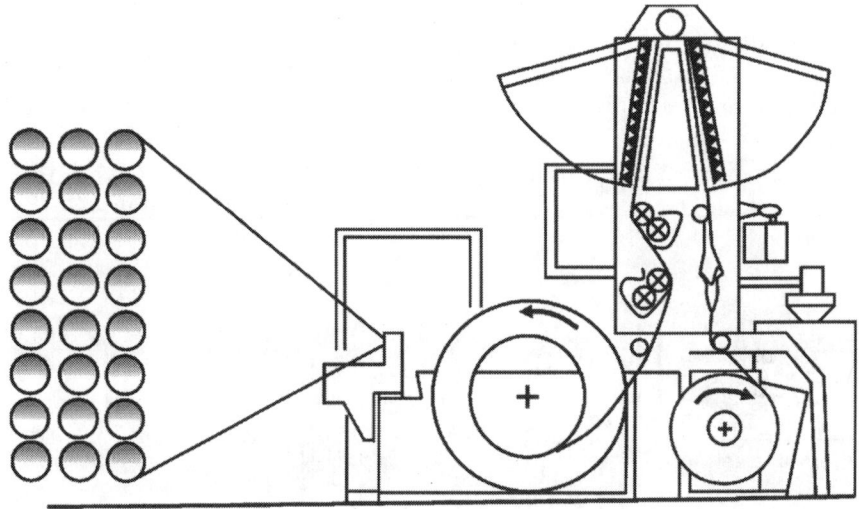

Fig. 4.71 Cold sizing

- High and uniform drying performance
- Fast reaction time when switching the radiators on and off, hence no over drying at machine stop or at low speed
- Precise temperature control
- Contact-less drying promotes the smoothening effect of the warp threads by applicator rolls.

Advantages of this sizing process are
- Accurate and controlled liquor application
- Gentle sizing of warp and virtually without loss in elongation
- Extremely high viscous and highly concentrated size could be used
- Savings in labor and space
- No size cooking required
- Capital expenditure for a lubrication device is low

After a long research and development work, Anjana Textile Ltd. developed a cold sizing process with two typical cold size recipe TEXO-9 and TEXO-99 in India. The TEXO-9 is basically used on sectional warping with oiling device without any heating arrangement. Manufacturers claimed that wide variety of yarns can be sized with TEXO-9. This process is offered for protective sizing, surface coating, energy saving, less pollution, as well as flexibility, and low cost. Cold sizing chemical TEXO-99 is able to size twisted, intermingled polyester filament, nylon filament, viscose and pure silk yarn, glass fiber, spun yarn, single cotton yarn, tire cord yarn with all possible twist multipliers.

HFZ combined warping-sizing machine
This machine combines sizing and warping in one machine, suitable for small orders, multi-yarn varieties, fine-count high-density weaving lines of cotton, dyed yarns, blended fabrics, linen as well as industrial fabrics, especially for small yarn dyed orders. This machine is useful to draw the warp thread layers from warping creels to size box. HFZ combined warping-sizing machine is shown in Fig. 4.72.

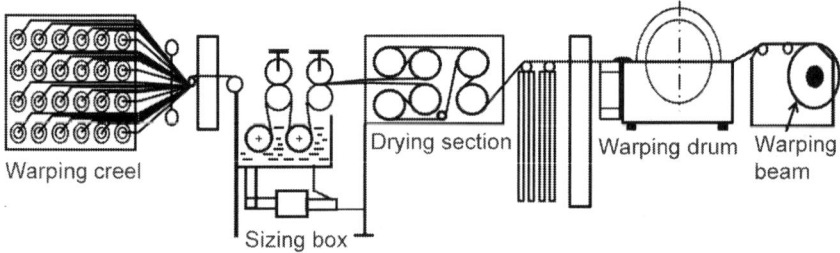

Fig. 4.72 HFZ combined warping-sizing machine

This machine is useful to process a wide range of warp threads used for warping and sizing. It has the facility to manufacture beam diameter up to 1000 mm with working width of 2300–3600 mm. The 10–80 m/min sizing speed can be achieved on this machine with beaming speed of 0–100 m/min. It has an effective brake system with less than 2 m accuracy. It has endless section traverse facilities with traverse speed of 0.001–9.999 mm/rev. It can minimize section positioning error up to ±0.01 mm. Distance between adjustable reed and drum remains constant. This machine provides automatic leasing facility. Six-drum drying system with steaming force of ≤0.35 mpa is used on this machine. The sizing box has the capacity of 80 liters.

4.33   Assessment of quality and performance of sized yarn

The grey yarn obtained from yarn spinning department has to subject under different preparatory processes like winding, warping, and sizing before fabric formation and the yarn has to withstand various kinds of stresses imposed by different mechanisms. The single yarns can seldom be woven in grey state because they have insufficient abrasion resistance. The generation of hairiness is the main culprit to the weaving operation. The type and quality of the grey single yarn has decisive effect on the performance of yarn in weaving apart from abrasion resistance and hairiness. A weak, fuzzy and non-uniform yarn will break frequently; whereas a smooth, uniform and strong yarn will be able to withstand the weaving stresses (Azarchab et al. 1982). According to Iyer, assessment of yarn in terms of count, strength, evenness, variability in count, and strength is not enough. The yarn defects like knots, gross faults, hairiness crackers, snarls, etc., are also equally responsible for poor yarn performance. For efficient and successful weaving, the yarn must have maximum possible elongation-at-break along with strength (Schwab 1980). For satisfactory weaving, abrasion resistance and hairiness-free yarn are prime prerequisites for satisfactory weaving of yarn. This weakness of grey yarn can be overcome by sizing process. Hence, some researchers agreed that the prime objective of sizing is to impart a judiciously thick band of size film around each yarn and cement the fibers inside the yarn together to protect the yarn from abrasion during weaving. The fixation of fibers inside the yarn prevents the inter-fiber slippage. Consequently, as more number of fibers starts to contribute in yarn strength, it goes up. The protruding fiber ends are bound together for their judicious utilization in yarn system. Thus laying of fibers results in improved compactness and smoothness of yarn. An optimum penetration of the size would be desirable so that all fibers are not either lying on outer surface of the yarn or glued up to the core of yarn.

Farrow and Hall (1980) revealed that the prime object of warp sizing is to produce weavability of the warp yarn or to overcome the inherent deficiencies of yarn. In sizing, the size must produce two fundamental properties in the yarn to achieve improved weavability: (i) Prevent excessive clinging and fuzzing by abrasion of yarn with itself and with loom parts (ii) Adequate fiber to fiber bonding to produce sufficient tensile strength in the yarn to protect its weak areas.

Houghton (1982) comments that the fundamental purpose of sizing of any type of warp yarn is to provide one or more of the following requirements, i.e., weavability, weight, and given finish to ensure higher weave room productivity. Textile research institute comments that "weaving coast can be increased or decreased by at least 25%, depending upon whether the warp sizing or dressing has been performed properly or not."

Slauson et al. (1985) revealed that sizing is a coating process in which polymeric film-forming agents are applied to warp yarns in order to make them appropriate to withstand with tensile, bending, and abrasive stresses that are applied during fabric manufacturing.

4.34   Size add-on % on warp yarn

Size add-on% on warp yarn depends on many parameters as shown in Fig. 4.73.

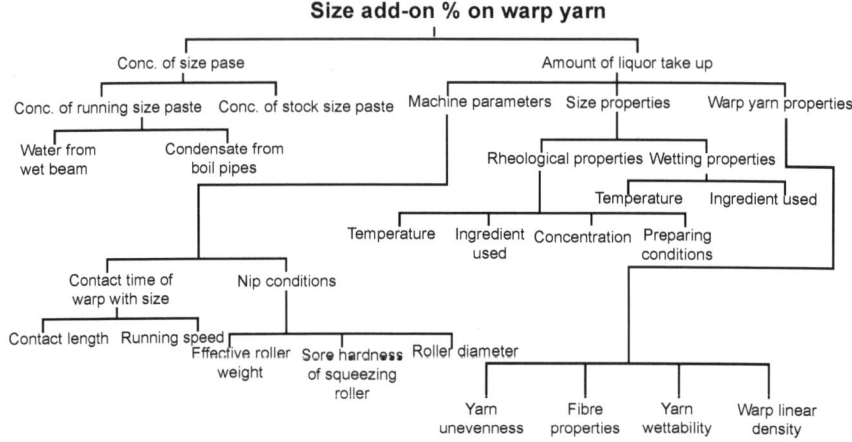

Fig. 4.73 Factors affecting size add-on

4.35   Factors affecting the quality and performance of the sized yarns

The increasing production rate of latest looms has made the sizing operation more critical.

The optimization of size pick-up can be defined by the substance balance that enters and exits the size box. To keep size pick-up constant, it is essential to maintain the temperature and sizing concentration in the size box as well as keeping the squeezing pressure and sizing speed constant throughout the process. Moisture content in warp thread at the size box exit and pressure of last squeezing rollers are the final process parameters that control the size pick-up. Other parameters like size temperature, sizing speed, thread tension, viscosity of size, levels and circulation of size in the box, sizing agents, conditions of size preparation and yarn properties also have remarkable impact on size pick-up (Goswami et al. 2004; Kovačević et al. 2004; Soliman 1995; Pleva and Rieger 1992). The size pick-up can be calculated by using the following relationship.

$$S_p = \frac{W_{S_p} - W_H}{\dfrac{100}{C} - 1 - \dfrac{W_{S_p}}{100}} \; (\%)$$

Where S_pS_p is the size pick-up, W_HW_H warp moisture at the box entry in %, WS_pS_p warp moisture at the box exit in %, and C is the size concentration in the box in %.

Homogeneous sizing ensures very smooth weaving at higher productivity and lower cost. It has been proved that the optimal size pick-up is the one that will allow minimum warp thread breaks on the weaving machine as well as satisfactory efficiency of the weaving machine and fabric quality. It is clearly mentioned in Fig. 4.74 that the optimal size pickup is slightly higher than the minimum, because it is very difficult to maintain the precision and uniformity of size pick-up especially on single cotton yarn.

The performance of sizing process depends upon numerous factors which are given below.

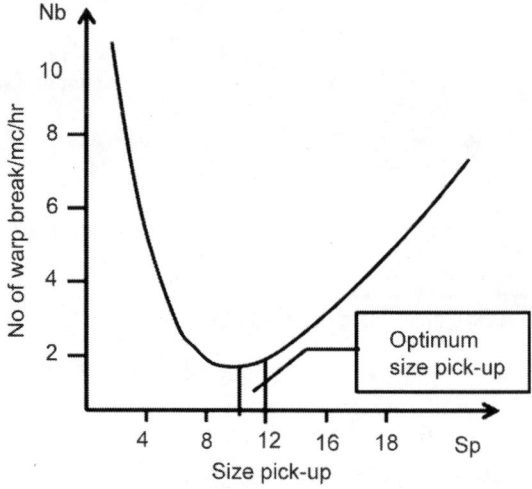

Fig. 4.74 Size pick-up

4.36   Online size add-on measurement by size application meter AS120

Pleva GmbH Germany developed an online size add-on measuring system with a commercial name of "size application meter AS120". This tester is able to do a contact-free, non-destructive, and most importantly, real-time measurement. The application of AS120 is very simple. It is directly placed between sizing box and drying zone. The measurement preciseness of AS120 does not get influenced by splash water, vapor, and temperature fluctuations like disturbances. In AS120, a temperature controlled hot air-flow through the measuring frame and heads prevent the influence of steam vapor on the measurement after the sizing box, as well as condensation on the frame. Homogeneous conditions are maintained by the continuous heating-air-flow which creates a warm air cushion between the measuring heads.

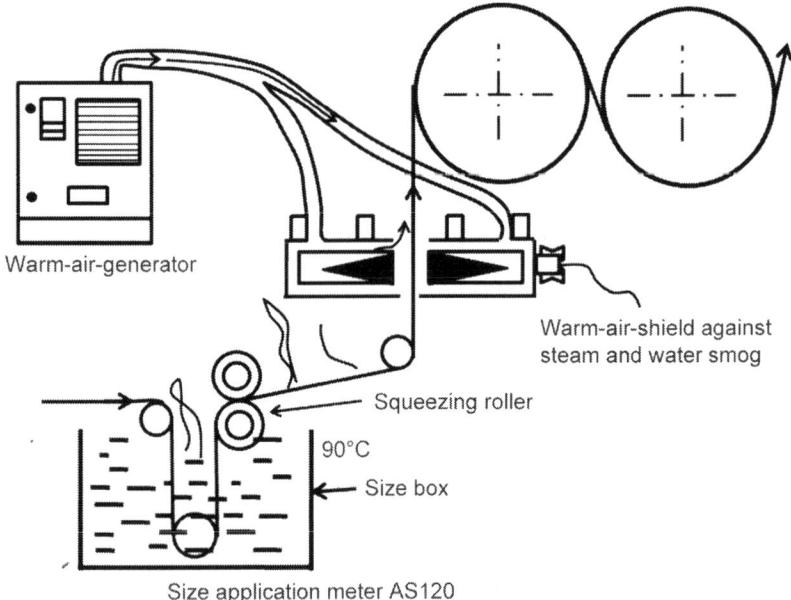

Size application meter AS120

Fig. 4.75 Size application meter AS120

Sized yarn runs between the two measuring heads without contact which makes this technique more precise. The contact-free measurement becomes possible by the application of microwave (microwave is absorbed by water molecules). The continuous microwaves sent by microwave generator are absorbed by the amount of water present between the measuring heads and the unused microwave signals that are received and analyzed at the receiving end.

Measured signal indicates the absolute amount of moisture on finely divisional scale. The lower the moisture content, the smaller the displayed value in scale divisions. For a constant size concentration, a change in the moisture content, respectively the displayed value corresponds to a change in the degree of size add-on. In principle, moisture content and size properties add-on to each other for a constant size concentration and the degree of sizing is accessible. In this way, a contact free and continuous assessment of the size add-on is possible in any case as shown in Fig. 4.75.

The optimum degree of size add-on depends upon parameters such as fiber type, applied sizing recipe, and so on. AS120 becomes very helpful to ensure the reproducibility of a sized yarn for a long time. The necessary pressure of the squeeze rolls can be adjusted exactly and instantaneously to exhibit the impact on the degree of sizing by means of the AS120. As a result, the AS120 offers the possibility to analyze and optimize the degree of sizing upon changing from normal production speed to very slow motion. Application of microwave concept in AS120 assures the quickest possible signal-processing and display, and therefore, provides an opportunity to adjust the degree of sizing to the optimal, lowest possible degree of sizing. Consequently, the consumption of sizing agent is significantly minimized because the undue application of size for safety and environmental reasons are no longer essential for the constant control of the size add-on. The application of AS120 helps in a uniform degree of sizing, thereby stabilizing the weaving efficiency on a high level by efficient measurement and effective control. This system is very useful in case of double size-box sizing system.

4.37 Size material

Sizing process is dedicated to the impregnation of warp yarns with a film-forming adhesive, to reinforce the warp yarns, for improved weavability. Hence, the adhesive behavior of film of binders with fibers must be studied properly.

Thus, a satisfactory warp-size paste must be proved appropriate on the basis of two factors.

(a) *Paste characteristic* properties which govern the behavior of size during preparation and application:
- Homogeneous viscosity
- Ease of preparation
- The pH near neutral point (7.0)
- Absence of foaming properties
- Absence of prolonged tackiness during drying
- Compatibility with other components of size mixture
- Stability towards decomposition
- Odor

(b) *Size film characteristics:* Following properties of size film must be optimized to exhibit satisfactory performance
- Adhesion to the fiber and yarn
- Tensile strength of size film
- Hardness of size film, especial resistance to abrasion
- Flexibility and folding ease of size film
- Insensitivity of size film to over drying
- Moderate insensitivity to change in relative humidity
- Little tendency to develop static electrical charges on abrasion
- Ready resolubility regardless of the age of the size film for ease of desizing
- Non-abrasive and non-corrosive to machine parts

According to Mousa (1978), the most important characteristics and those which contribute most heavily to the weavability in sized yarn are adhesion to fibrous substrate, elongation, elasticity, flexibility, resistance to abrasion, viscosity, and desizability of size film. Out of these, two factors appear critical to size performance

(a) The size must form a tough but flexible film to protect the yarn from flexing and rubbing actions

(b) The film forming properties of the size provide some insights into abrasive resistance of the sized yarn.

According to Bancroft (1934), generally the adhesion depends on adsorption of the adhesion by the surfaces to be the joined, and that the degree of adhesion can probably be measured by the amount of absorption.

(c) *Effect of size add-on*

Apart from the physico-chemical properties of the sizing material, the quantity of the size applied onto the yarn is of vital importance in assessing the quality and performance of the sized yarn. The size add-on is governed by a number of factors such as machine speed, squeeze pressure, viscosity, and temperature of size paste. An increase in speed and concentration increases the size add-on, while an increase in squeeze pressure decreases the size pitch-up. It is evident from series of experiments that sizing improves yarn strength and gain of strength increases with increase in size add-on. At very high add-on values the yarn becomes stiffer and less flexible because of a thick shell of size encapsulating the yarn and a deeper penetration.

- Size add-on below or above optimal value increases the number of warp breakages
- Over sizing reflects higher cost through higher size consumption, large liquor volume, higher drying energy demand, and higher effluent levels
- With increase of size add-on, the tenacity increases but the extensibility decreases

Various researchers described the sizing-weaving curve by plotting the add-on vs. warp breakage for different size materials under different conditions. It is found that ultimately weaving is possible with much difficulty in case of unsized yarn. After this, less and less trouble comes and warp reaches at its best weaving state. Further increase of size add-on on warp yarn may have little or no effect but ultimately, yarn becomes stiffer and weaving again becomes more difficult.

Kovacevic et al. (2008) studied the impact of various colors in a multi-colored warp on size pick-up in order to analyze the change in physical-mechanical properties of yarn after sizing. They concluded that various color warps were exhibited significant difference in size pick-up and physical-mechanical properties after sizing.

First, the warp sheet is immersed in the size liquor, and afterwards it is squeezed and dried to remove the moisture that is often lower than the one that it had prior to sizing. Sizing is the most expensive stage in weaving-preparatory section due to very high consumption of energy (steam, and electrical power), sizing agents, and water. Integration of various software-supported measuring devices make the continuous tracing and controlling of various process parameters influencing sizing quality (Exbrayat 1992; Fiebig et al. 1995; Pleva et al. 1992; Schalze et al. 1993; Soliman 1995). Majority of researchers agreed that it is very difficult to maintain the size pick-up constant. Although, the sizing conditions affect size pick-up, the interspaces among the fibers in yarn body also play important role in deciding the size pick-up.

(d) *Desizing*

Size application is a temporary application for improving the mechanical properties for efficient weaving and the yarn has to be desized for most of the fabrics before wet processing of it. The size ingredients should be easily and efficiently removed from the sized yarn. The degree of size removal depends essentially upon the following factors:

- Size coating/penetration affected by viscosity of size paste
- The size add-on
- The fabric construction parameters and fabric set
- The resolubility of the size film
- The desizing process

Another important aspect of size materials is the extent of pollution caused to the streams, river water, etc., by the effluent discharge from sizing and desizing sections of textile industries. Biochemical oxygen demand (BOD) for the adhesives is in the following order:

Starch > modified starch > synthetic sizes

But the chemical oxygen demand follows the reverse order.

4.38   Effect of pre-wetting of warp yarns

Sejri et al. (2008) performed a series of experiments to establish the benefits of pre-wetting of warp yarns before being immersed in size paste. They revealed that the heat-transfer coefficient was remained almost constant in case of both non pre-wetted and pre wetted yarns without any change in productivity. However, despite containing less size, the yarn hairiness was reduced as compared to sizing without pre-wetting. The initial modulus of sized yarn with pre-wetting was higher than that of conventionally sized yarn, and the sized yarn with this new concept of pre-wetting of sizing offered a more important immediate recovery. This study revealed that the influence of a preliminary wetting on the structure of the size bridges between the fibers. They confirmed by SEM study of the cross-section of sized yarn that if the yarn was pre-wetted, the bridge frequencies between fibers were enhanced to a higher extent at the yarn periphery, which promised the improved mechanical performance during weaving.

Schwarz et al. (2011) have made constructive efforts to make a comparative analysis of the standard sizing process and pre-wet sizing process. They found lower consumption of sizing agents and power with pre-wetting sizing concept. It is well established in textile industry that the upgradation of the conventional process and installing a part of the sizing machines necessary for the attachment of the pre-wet facility does not require a large technical and financial burden because sizing with pre-wetting arrangement gives superior results. They inferred that the properties of yarn sized with recipes of a lower size concentration (5%) showed very good results in terms of insignificant deviations (despite a much smaller amount of size pick-up on the yarn) as compared to the recipe with a higher-size concentration (7.5%), especially in the pre-wet sizing process, in which some properties are even better. It is evident from the above findings that pre-wetting process opens new possibilities to reduce amount of sizing ingredients, water, and energy costs (both for the sizing and desizing process) without any negative impact on the sizing process and properties of sized yarns.

4.39   Influence of yarn quality and fiber blend

The use of synthetic fiber has made the sizing process more crucial. The selection of size recipe is mostly influenced by fiber blend and yarn structure.

4.40   Mechanical and physical properties of sized yarn

Langer (1985) described that the following parameters of sized yarn are key parameters in order to evaluate their weavability.

(a) *Breaking strength*
Sizing improves the yarn strength. Above a particular level of yarn strength the warp-breakage rate is insensitive to increase in strength, but with a reduction in strength below that level the warp-breakage rate increases sharply.

(b) *Breaking extension*
The process of sizing reduces the breaking elongation of grey yarn. According to Slauson et al. (1985), this loss in yarn elongation during sizing takes place because of following factors:
(1) The cementing of the fibers by size paste which precludes the inter-fiber slippage
(2) The tension applied on warp sheet during warping and sizing reduces the residual yarn extensibility
(3) The yarn drying conditions reduce the yarn extensibility
(4) The yarn elongation should not be reduced more than 20–25% by the sizing process.

   Few studies have shown that a less extensible yarn cannot withstand, for a long time, the cyclic stresses and abrasion on the loom and therefore will lead to breakage.

   Slauson (1984) concluded that although the yarn strength increases on sizing, probably due to increase in yarn cohesiveness and prevention of inter-fiber slippage, yarn tenacity probably (above an optimum level) is not a critical property by knowing the fact that the tensile force exerted on the yarn during weaving is only 20% of its breaking strength. On the other hand, yarn extensibility looks to be more critical if the yarn is to withstand the cyclic extension and abrasion, bending and flexing action of the loom.

4.41   Sonic modulus

Sonic pulses travel fast when the material through which it is propagating is more oriented, crystalline, compact, and with minimum flaws. As the sonic pulses move faster through material, the sonic modulus gets higher. Sonic modulus represents the modulus of the material under dynamic high frequency conditions at very low strain level. Some researchers floated the views that the sonic modulus is more correlated to weavability than the initial modulus.

4.42   Work of rupture

The work of rupture indicates the toughness or hardness of the material. The area under the load elongation curve is known as the work of rupture. According to Morton and Hearle (1962), work of rupture is the appropriate quantity to correlate with the potential of bear sudden shocks by yarn (like beating up, shedding stress, etc.).

4.43   Hairiness

Yarn hairiness is an important criterion in weaving because the protruding fibers of neighboring ends cling to each other. Sizing reduces hairiness to a major extent by providing surface coating to the yarn.

Trauter (1980) concluded that the yarn hairiness increases with the decrease of add-on and increase of residual water content up to 10%.

Different researchers reported about the effect of size add-on on the abrasion resistance and found that the abrasion resistance increases with the increase in size add-on on yarn (Aggarwal et al. 1983; Slauson et al. 1985). However, the rate of increase depends on the yarn quality, abrasion device used, and type of sizing agent. It is based on the fact that the number of necessary cycles to rupture the yarn at low add-on is less because the penetration and the size shell encapsulating the yarn is not as it is at higher add-on.

4.44   Yarn structure

The staple yarn geometry mainly depends on fiber and yarn fineness and amount of twist in yarn to influence hand, drape, abrasion resistance, and strength of fabrics. Most of the geometrical parameters of a yarn like packing density, yarn diameter, density variation, and circularity of the yarn along with surface characteristics are affected by the yarn twist. Sizing plays an important role in deciding the yarn structure before weaving.

Schwarz et al. (2011) studied about the changes in mechanical and deformation behavior of single cotton yarn after sizing, and concluded that the mechanical and deformation properties depend upon fineness, size concentration, and the structure created by the type of yarn spinning. The cyclic testing to determine areas of elastic, viscoelastic, and plastic yarn deformations of ring and rotor-spun yarns were analyzed before and after sizing with a PVA sizing agent, using the same recipe, but with two different size concentrations of 7.5% and 5.0%. The study did not show any remarkable difference in yarn properties after sizing. This point draws a conclusion that it is a greater advantage to use a lower concentration to achieve better economic and environmental effects. The improvement in ring-spun yarns was more noticeable in comparison of rotor-spun yarn due to its typical structure. The yarn structure is a very important parameter to a difference in changes of deformation areas of unsized and sized yarns, primarily in the way of reducing the elastic area of deformation. The major inferences of this study are given below.

- The absorptive properties of the ring-spun yarn are much smaller than the rotor yarn, which affect the amount of size pick-up on the yarn. The difference in absorption potential of ring and rotor yarn depend on geometry and extremely highly interconnected fibers within the yarn, the ring-spun yarn is much stronger, with greater breaking force, yet with much less elongation than the rotor yarn, whose geometric arrangement and connections of fibers are not as strong.
- Differences in breaking force and breaking strength of the two yarn types are even more accentuated after sizing, where the sized rotor-spun yarn does not experience big changes after sizing.
- The unsized ring-spun yarn has twice as many protruding fibers as the rotor-spun yarn but this gigantic difference is completely diminished by sizing, after which the number of protruding fibers is almost equal.

4.45   Factors affecting the size yarn performance

Recent developments in high speed weaving machines have increased the pressure on sizing process critically. The effectiveness of the sizing process depends not only on the adhesion between size and yarn but also on film forming and rheological properties of the size, process variables, and yarn properties.

4.46   Size material

The effectiveness of size materials depends on the type of material and mass applied. The advance research in chemistry has brought awesome changes in chemistry of size materials and altered the situation of size materials fundamentally. Now a day, a very wide range of sizing materials is available which act as film forming materials, binding the protruding fibers within the warp yarn body, and imparting enhanced strength, elasticity and lubrication for high production weaving machines. Conventionally, starches have been used for sizing of spun cotton yarns by aqueous process.

4.47   Effect of drying system

Katovic et al. (2008) have studied the effect of various drying methods on physical-mechanical properties of yarn. The study of effect of drying on breaking strength reveals that microwave drying (MW drying) system is best for the drying of single yarn after impregnation in size box. The maximum elongation at break is found in case of contact drying on conduction principle, but microwave drying proved to be even better when minimal pressure force was applied. The abrasion resistance analysis showed that microwave drying system provided a slightly higher abrasion resistance under all drying conditions. The yarn hairiness of microwave-dried yarn was lower than the convection-dried yarn and higher than that of the contact-dried yarn. The microwave drying does not make any surface contact with drying surfaces and saves the cross-sectional shape of yarn after drying. The microwave drying system enhances the sizing speed remarkably.

4.48   Features of new sizing methods

Industrial, environmental, and economic forces always motivate the machine manufactures to incorporate new features in latest models of their machines. Few features of new sizing machines and methods are listed below.

- Low energy consumption
- Size cooking consumes lot of energy, which can be saved if cooking can be avoided or made more efficient
- Normally, 70–80% of total energy consumption is required for drying of sized wet yarn. New methods save energy during drying.
- Low sizing material requirement with improved size material in terms of viscosity, concentration, and adhesive power, and the sizing material requirement is less. Moreover since size material cost is the main component of total sizing cost, any saving in sizing material saves a lot of money.
- Better sizing quality is achieved through improved size-application technique and drying.
- Less space requirement with hot melt and cold sizing
- No separate machine for sizing is required, as the process is incorporated with warping. Moreover, space for size cooking is also not required in any of these methods, i.e., hot melt, solvent and cold sizing.
- Increased sizing speed. In applications where the sizing speed is not limited by drying, sizing speed goes up considerably.
- Reduced workload on operators
- The new methods are less complicated, easy to operate, and are user friendly.

4.49   Latest developments in yarn sizing machines

Various renowned sizing machine manufacturers have made remarkable contributions to their sizing machines in order to enhance better quality production with least damage to the environment. Toyoda Machines, Japan, Tsudakoma Machinery, Nomachi Kanazawa, Japan, KARLMAYER, etc., are the major leaders in sizing machine manufacturing globally.

4.50   Contribution of Toyoda Machinery, Japan

Toyoda Machinery, Japan, manufactured sizing machines of various generations which are discussed in many old books. The contribution of Toyoda is discussed here in terms of their latest sizing machine Filamaster Express 610. The features of Filamaster Express 610 are listed below.

4.50.1   Filamaster Express 610

Filamaster Express 610 is a pioneer sizing machine of new era. This sizing machine is able to produce superior sizing quality at high speed and maintains homogeneous yarn tension throughout the process. This sizing machine is equipped with following features:

Improved quality of size paste
With improved circulation of the sizing liquor that overflows from the front and rear of the size box and a higher quality sizing liquid filter from the cavity box, bubbles are effectively prevented in the sizing liquid during high-speed operation, which can adversely affect the sizing quality. During high-speed operation, the sizing liquid level is lowered to prevent splattering. Two different circulation routes of sizing liquid remove impurities from the bottom to ensure a constant flow of pure and high-quality sizing liquid to the size boxes.

Hot-air drying
Hot air is applied in and against the warp yarn sheet's running direction, and there are two continuous drying chambers, resulting in much higher drying capacity for the warp yarn sheet.

Uniform drying
The automatic temperature control system using motor valves adjusts the volume of steam and maintains a fixed temperature in the hot air chamber to ensure uniform drying of the warp yarn sheet regardless of the operation speed.

Wet splitting
A water-cooled wet dividing rod divides the warp yarn sheet immediately after sizing. The moisture on the rod surface prevents size waste from accumulating on it.

Homogeneous yarn tension throughout the process
The tension of the warp yarn sheet is controlled in real time by the AC vector inverters which work in the let-off and take-up sections. The machine's superb response with digital control provides precise operation at normal speeds as

well as during machine acceleration, deceleration, and stoppage, thus assuring stable yarn tension.

Digital control of yarn stretching
To ensure optimal stretch to the warp yarns according to the yarn type and maintain superb quality yarn during the drying process, the Filamaster Express 610 adjusts the revolution speed of the squeezing roller and drying cylinder. The machine measures the rollers' revolution speed and indicates the stretch in terms of the rollers' revolution ratio to control the residual extensibility in sized yarns.

Heating of hot air chamber
The heat in the hot air chamber is generated by a conventional steam heater as well as an electric heater. With this high drying ability, the temperature can be quickly increased as soon as the machine starts and during ultra high-speed operation. Per-block temperature control inside the cylinders maintains the quality of the warp yarn sheet.

Uniform squeezing of warp sheet
A specially designed EP roller (equal pressure roller) squeezes the center and edges of the warp yarn sheet uniformly, preventing irregular sizing of the warp yarn sheet, even during high-pressure squeezing. Synchronized control of the machine speed and the squeezing pressure enables the squeezing pressure to adjust according to any changes in operation speed, meaning stable sizing even during machine acceleration and deceleration.

Easy operation
The Filamaster Express 610 is equipped with a user-friendly touch-screen function panel. With the touch of a finger, the screen displays operating conditions such as take-up length, speed, and tension, allowing the operator to quickly confirm and adjust the settings.

Problem location finder
The machine stops automatically if a problem occurs during operation and shows the exact location of the problem on the monitoring display.

Exact moisture content in sized yarn
The temperature indicator records the temperature of six different points from among the size box, hot air chamber, and drying cylinder sections in a graph, giving the operator complete control over the quality of sizing. The circulation fan inside the chambers uses an inverter control system, which can automatically set the fan speed to match the operating conditions, to provide consistent, efficient drying, and reduced energy consumption.

Operator safety

The Filamaster Express 610 includes various safety features for the operator such as covers over the high-pressure cylinders, which can become very hot during operation, as well as covers over the drive section and rollers, and an emergency stop button on the main unit.

4.51   Contribution of Tsudakoma

Tsudakoma Machinery, Nomachi Kanazawa, Japan has launched a number of latest sizing machines of KSH series (KSH 300, KSH400, KSH500, and KSH 600) for filament yarn sizing with following features:

(a) Uniform squeeze of sized yarn for ideal sizing

In the case of KSH series high-speed sizing machine, the squeeze roll achieves even widthwise sizing constantly in order to achieve homogeneous sized beams. A high pressure squeeze system of up to 15kN is adopted. A smooth, stepless pressure-control system works with yarn speed automatically.

(b) The ideal roll arrangement inside the size box with large capacity of 180 l prevents size splashing and foaming.

The cylinder drying and the waxing/cooling device make high speed operation possible

The KSH400/500 is equipped with five cylinders (three cylinders for the KSH300), and seven cylinders are available for high speed thick yarn sizing. The heat-resistant oil sealed chain is used to drive the cylinders. The yarn breakage sensor attached after the cooling device is used as a part of thread-stop motion that can automatically stop the machine.

(c) Multi-section drive system

Various types of multifilament yarns like twisting, non-twisting or texturizing are processed on sizing machine which needs frequent changes in stretch setting to adjust the draw ratio when replacing the one type multifilament yarn by another one. The AC vector motors of M1, M2, and M3 control the stretch for the feeding section, sizing section, and drying section respectively by 0.1% increments in draw ratio to control any slackness. This machine provides facility to set draw ratio very easily while the machine is running.

(d) Sensitive and precise temperature control

Preciseness is very essential in temperature, and tension in achieving the best quality sized beam. The KSH sizing machine series is equipped with precise temperature control at the hot air chambers, the individual cylinders, the waxing, and the size box. Current temperature is indicated and recorded, and irregular temperature sounds the alarm to stop the machine.

(e) Energy saving hot-air chambers for high speed

The hot-air drying chambers are available in this machine to maintain the surfaces of the yarns smooth and the body round without touching any surface during drying. The chambers of the KSH300/400/500 use the two-step temperature control system that keeps the temperature at 150–160°C for high speed operation and at 120–130°C for low speed operation. In this way, appropriate drying is achieved quickly without altering the basic yarn characteristics.

(f) Easy operation

Automatic warp beams mounting and unloading facility is available in creel section of KSH series machines. The KSH sizing machine is designed smartly for easy threading through the four yarn separators to arrange the yarn sheet at the important points on the machine. A centralized lubricating system is integrated with these machines that automatically lubricates to 100 or more positions in a set cycle to achieve homogeneous and precise lubrication.

(g) Simple and reliable computer control system

The machine data station (MDS) systematically controls the KSH machine operations with the help of touch-screen operation on the display and colored graphic display. This arrangement discards the possibility of any mis-operation.

The MDS-II is available as an option with an improved version of KSH500. The MDS-II collects production management data most effectively to assure highest quality and reports operational results, and it prints out size pick-up information from the size pick-up monitor (SPM).

(h) Movable tensioner frame (for spindle pitch of 310 mm or more)

An electric-drive system is integrated to shift the tensioner frame to align the centers of the warp beams and tensioners. Generally package shapes vary according to yarn kinds, and this system helps to adjust the tension variation. The sizing efficiency of machine is improved by saving package holder exchanging time.

(i) Minimum tension variation

The tension variation among the yarns is minimized by assuring the small yarn bending angle because sharp change in yarn bend angles creates remarkable amount of tension variation.

(j) Stabilized take-up control for fine to thick yarns

A wide range of yarns from fine to coarse are controlled to minimize the tension variation by the application of load cell, speed controller and tension controller as shown in Fig. 4.76. Speed meter and tension meter are integrated with speed controller and tension controller for very fine tuning automatically during running.

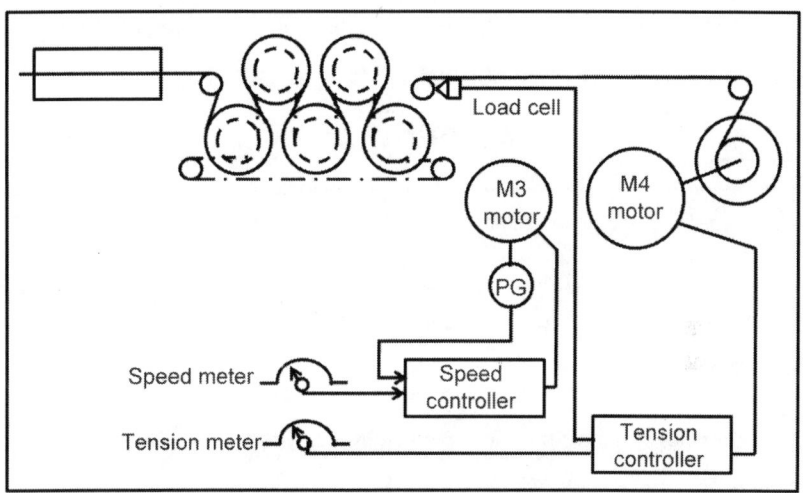

Fig. 4.76 Stabilized take-up control by Tsudakama

(k) Quick and easy doffing

The supply packages can easily be replaced from outside the creel, ensuring easy doffing. The next packages to be doffed are set on the outer spindles of the spindle frames. After completing one set operation, spindles on each semi-magazine frame are turned 180° in a unit of 24–32 spindles with a pedal. Supply packages become ready for the next operation in a very short time.

(l) Middle guide

In creel zone, the yarn interaction with each other is prevented by the use of middle guides that are made of alumina porcelain. When spindle pitch is 300 mm or more, and the number of rows is numerous, the middle guide is divided into two steps making its pitch about 40 mm. Additionally, the optional static-electricity eliminator is mounted on the middle guide section to reduce static electricity if required.

(m) AP automatic kink preventer (optional)

A kink preventer is necessary to draw twisted yarns from the creel. The AP automatic kink preventer protects twisted yarns from forming kinks with tricot knit fabrics. The tricot knit fabrics hang on both sides of the supply packages, and simultaneously come close to the packages at the push of a button.

(1) Heat cutter (optional)

The heat cutter simultaneously cuts yarns along a guide rail on the floor. Since yarns are cut near supply packages, enough length for tying remains. This system is very much suitable for synthetic filament yarn. The cutter uses a battery with charging equipment.

Table 4.2 Differences between KSH 300, KSH 400, and KSH 500 series.

	KSH 300	KSH 400	KSH 500
Maximum yarn speed m/min	300	400	580
Maximum sizing and drying width	1700 mm	1700 mm	1700 mm
Take-up width, ( ) means a special design	1700 (1780) mm	1780 (2000) mm	1780 (2000) mm
Sizing system	Double squeeze	Double squeeze	Double squeeze
Outer diameter of immersion roller	ϕ 180 mm	ϕ 180 mm	ϕ 180 mm
Outer diameter of sizing roller	ϕ 200 mm	ϕ 200 mm	ϕ 200 mm
Outer diameter of squeeze roller	ϕ 190 mm	ϕ 190 mm	ϕ 190 mm
Squeeze roller structure	Uniform squeeze	Uniform squeeze	Uniform squeeze
Maximum squeeze pressure	8 kN (kilonewton)	8 kN	15 kN
Drying system	Hot air drying	Hot air drying	Hot air drying
No of drying cylinder (ϕ = 800 mm)	3	5(7 for thicker yarns)	7
Take-up tension control system	AC vector tension feedback system	AC vector tension feedback system	AC vector tension feedback system
Take up tension control range	80–500 N 100–800 N	80–500 N 100–800 N	80–500 N 100–800 N
Outer diameter of applicable flange diameter	620, 800 mm	800, 1000 mm	1000 mm
Computer control system (MDS-II; optional)	MDS-I touch key operation	MDS-I touch key operation	MDS-I touch key operation

4.52   Contribution of some Indian sizing machine manufacturers

4.52.1   Indotex sizing machine with PLC control AC drive sizing machine with double sow-box (model Id-0005H)

Ahmedabad, India based manufacturer presented a sizing machine to fulfill the requirement of Indian customers in Indian working conditions. This model is totally computerized and specially designed for composite textile unit for having fully modernized technology loom creel, double sow box, two sets of pre-dryer and final dryer, head stock, and fully computerized AC frequency drive with all necessary electronic instruments and controls. Some silent features of this sizing machine are as follows:

- Great improvement in operation
- Available up to 3600 mm working width and 1250 mm beam working
- Even size pick-up
- Comprehensive PLC control AC invertors drive
- Precise starch and tension control
- Simple AC invertors and DC hyperbolic drive
- Available different models for different quality of fabrics
- Custom design to produce high-quality size beam at low cost to achieve highest weaving efficiently

4.53   Developments by Toyoda Machinery, Japan

The higher speed of air-jet and rapier looms forced to develop better sized yarn having fluff in the warp sheet be firmly laid down and that the warp sheet be uniformly sized and dried. To fulfill these requirements, Toyoda Machinery Japan introduced the Mackee Eagle spun yarn sizer, designed and developed to control the stretch and drying of sized yarn very precisely.

The creel section of Mackee Eagle is equipped with electromagnetic convertor to control the over stretching of warp yarns without allowing any slackness in warp sheet. Toyoda's Mackee Eagle sizing machine is shown in Fig. 4.77. The electromagnetic converter block provides a generic interface between the electrical and magnetic domains. An electromagnetic converter includes a magnetic circuit having permanent magnets. Warp beams are arranged in double-dicker system in creel section. Warp threads from both rows of beams are sized in two different size boxes. Top row of warp threads is sized in first size box and bottom row in the second. After getting size in both saw boxes, both warp sheets are dried separately in drying chambers situated just ahead of both saw boxes. After that both warp sheets are combined and dried simultaneously in third drying chamber. Dried warp sheet passes under a cushion roll and couple of other rollers which are attached with tension and stretch sensors. Finally, the sized warp sheet is wrapped on weavers beam.

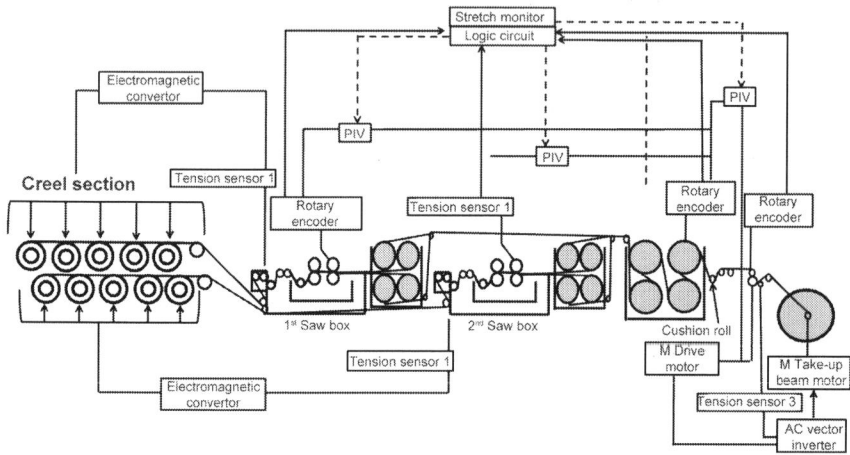

Fig. 4.77 Toyoda's spun yarn sizer 'Mackee Eagle'

4.53.1   Tension control optimized for multi-variety, mixed yarns

The Mackee Eagle takes each group of yarns as a batch and measures sheet tension at the front of each size box. According to the measured tension, it automatically distributes control of yarn tension on the beam stand by combining band brakes with pneumatic pressure. Additionally, when simultaneously sizing beams with extremely different yarn densities or yarn types, such as dyed beams, the Mackee Eagle is optionally provided with a device that enables automatic tension control to be disengaged to allow manual tension adjustment by means of proportional integral velocity (PIV) controllers. This versatile tension control system is designed to handle the widest possible range of today's increasingly diverse warp ranges.

4.53.2   Reliable design eliminates drying defects and over drying

The manufacturer has freedom to choose the optimum cylinder temperature control system to match the type of warp sheet being sized. A set of four cylinders is used to control the temperature and manufacturer offer a two cylinder set is also available as an option. To prevent over drying and other defects, moisture content detectors are installed at the cylinder openings to automatically control operation speed, according to the degree of drying and yarn type.

4.53.3   Operating conditions set by "system master" input

Mackee Eagle's manufacturer has provided the facility to assign a code for each set of yarn parameters. These parameters are yarn count, length, yarn type, number of ends in warp sheet, and reed width of the beam to be sized, and store them in the system master of the machine along with the machine parameters. This data base becomes very useful in future when a warp beam of similar parameters is to be sized, the machine operator or department supervisor simply puts its code and the system master of Mackee Eagle automatically adjusts, monitors, and controls machine speed, cut-marking, winding tension, temperature, squeezing settings, and other parameters, eliminating variations from lot–to-lot. This arrangement offers greater reproducibility which was very difficult in sizing before the invention of Mackee Eagle.

4.54   KARL MAYER

The KARL MAYER group is the new owner of the Sucker-Müller product section that has been acquired from the insolvent Moenus Textilmaschinenfabrik GmbH in Mönchengladbach.

The "Gebruder Sucker" factory was founded in Gebruder in 1881, and in 1983 it merged with the company Franz Muller under the name of "Sucker Muller". The fusion with the textile machinery manufacturing group "Hacoba" resulted in the "Sucker-Müller-Hacoba" group. After some years, the Sucker became the prime leader in warp preparation section and largest competitor of KARL MAYER. KARL MAYER purchased the Sucker-Müller-Hacoba group on November 13, 2007 and the complete purchase includes the trademark and patent rights, which enabled KARL MAYER to extend its warp preparation division by the internationally well accepted brand name "Sucker". This acquisition gives fresh impetus to KARL MAYER's further development in the growth sector related to warp preparation machines. Now KARLMAYER's flag is hosted at Obertshausen (Germany), Rotal (Italy), and Wujin (China).

KARLMAYER Textilmaschinenfabrik GmbH, Obertshausen, signed the contract on December 1st, 2008, to take over the weaving preparatory section of Benninger AG in Uzwil, Switzerland and announced that in future, it will work under the name of KARL MAYER Textilmaschinen AG. With this strategic acquisition, KARL MAYER takes over Benninger's current product portfolio, together with the service/components operations, which will remain in Uzwil. This takeover opened the ways for the emergence of a leading international company, offering an extensive range of products in the market for weaving preparation equipment and machinery. But other business operations of Benninger have remained unaffected by this takeover.

KARL MAYER aims to consolidate its worldwide service network by taking on Benninger's service personnel, who will continue to provide technical support for Benninger's installed machines in future as well.

4.54.1   Denim sizing with KARL MAYER – perfect solutions for weaving

Direct beaming and sizing technology has a decisive influence on the economic aspects and efficiency of the weaving preparatory section. A single machine manufacturer of direct beaming, dyeing, and sizing processes can be improved for user friendliness and efficiency.

The entire production chain in weaving preparatory section must be synchronized with weaving activities in order to optimise productivity of quality product.

For example, the sizing department must supply the weavers beam to the weaving department having the following characteristics:

- The least number of yarn breakages, and preferably no uniformly sized beams
- Uniform and targeted sizing of every yarn of warp sheet
- Constant residual yarn moisture content, without any stickiness
- No standing marks or marks caused by over-drying
- Constant yarn tension/elongation
- High yarn residual elongation
- Zero crossed yarns
- Low hairiness with no clinging
- Perfectly wound edges

To produce weaver's beam of above quality, the sizing department needs warper beams with the following:

- Least number of yarn breakages, and preferably no crossed yarns
- Constant winding densities and beam diameters
- Constant wound lengths
- Constant yarn tension
- Very little dust and fly, and preferably none at all

KARL MAYER's weaving preparatory machines are used in Denim sector and can meet all the demands optimally and reliably. To this end,

KARL MAYER has an experience of developing its machines for centuries almost, for all weaving preparatory operations. Consequently, the company has constantly been developing its machines for every processing stage. The journey of development was started from mechanical to electrical, and then electronic to computer controlled. A perfect integration is possible among the all KARL MAYER's machines. These modifications assure the accurate and uniform yarn transportation with extremely short and identical residual running lengths in every batch.

These high-tech machines have already achieved an extremely high degree of reproducibility at higher efficiency.

KARL MAYER group manufactured high efficiency warp preparation machines by using Kamcos as an essential accessory. Integration of KAMCOS® on KARL MAYER machines provided a standard platform for computer-controlled operations on sizing machines that offer the layout of intelligent networking concepts in textile industry in order to achieve the right conditions for carrying out modern, transparent production operations. The application of KAMCOS® control technology with KARLMAYER assures the reproducibility of a number of weaver's beam of same diameter, batch length, and yarn density. Wrappers beams having identical characteristics are a prerequisite for obtaining high-quality, consistent dyeing, and sizing.

The modern dye sizing concept ensures the judicious use of energy and the sizing ingredients in order to achieve enhanced productivity without sacrificing the product quality.

Sizing machines with latest technologies produce high-quality, uniform, reproducible warp beams that are needed to achieve the optimum weft insertion rates on modern, high-speed air-jet, rapier and shuttle looms.

4.54.2   BC-A beam creel and ABR controller on sizing machine

There are some factors which must be considered during quality control on sizing machine.

- The yarn tension must be kept constant at all times during unwinding of the heavy warper beams having flange diameters of up to 1400 mm.
- During machine downtimes, emergency stops, periods of acceleration, and also once the production speed has been reached, it is essential to avoid high tension peaks during the stoppages, together with the associated risk of over-stretching of the yarns at the warper beam.
- KARL MAYER machines equipped with KAMCOS® system, together with the associated data collecting system and an efficient

braking system eliminate any possibility of brake failure.

- A calculating system incorporated in the KAMCOS® system continuously supplies values on the required pneumatic braking system and stopping pressure for controlling the diameter of the beam currently being processed.

- Correspondingly, the yarn tension measuring rollers (load cells) in the ABR control system operate at a high resolution and with a high degree of precision. Flexible band brakes equalise any fluctuations between the actual and target values quickly and reliably.

- The band brakes are controlled by specially manufactured, low-friction roll diaphragm cylinders; they equalise out both resonance and beam irregularities and thus guarantee smooth machine running. This results in the shortest possible stopping times and distances, which avoid any further yarn breakages and yarn spirals.

4.54.3   Compact size box (CSB) and compact size box with pre-wetting (CSB-PW) sizing technology

The SMR sizing machine is PC-controlled and operates with the latest "triple dip and double nip" sizing technology in the form of the CSB application technology. The combination of the dipping/nipping technique, with highly turbulent application technology, creates a combined solution for the denim sector.

Initially two sizing systems had to be used when the yarn loading density was 70 or 80% higher, whereas now a days, just one CSB application system is able to size the textile material constantly and uniformly at yarn loading volumes of up to 110% smoothly.

Application baths having a low liquor volume and constant-pressure rollers operating at up to 40 kN pressure are used for sizing in CSB system. The CSB technology, which is constantly being improved and revamped to meet the changing technological, ergonomic and economic requirements of the market, was developed by KARL MAYER with the following latest features:

- An ergonomic design makes easier access for maintenance and cleaning the size box

- Having a liquor volume that is up to 60% lower than in conventional techniques

- The application of high-quality, corrosion-resistant materials bound textile transport during wet application coordinated rubber-coated rollers for avoiding differences in circumferential speed and resulting yarn displacement optimised squeezing and wetting technology for achieving higher yarn loading volumes.

CSB-PW wet-on-wet sizing technology

The wet on wet technique by pre-wetting of warp sheet with water involves the additive application of sizes onto warp materials.

The warp yarns run through a separate pre-wetting bath, integrated with high-pressure squeezing arrangement in order to absorb up to 40% warm or hot water. Water absorption mainly takes place by capillary action of yarns that prevents the size from penetrating into the interior of the yarn, provides a surface coating of size ingredients which can easily be washed out during scouring process. This machine integrated with a liquor consumption measuring system combined with an optional online microwave monitoring and control system, which avoids over-sizing and thus makes process economic. To optimize the further processes, degree of sizing and nip pressure is measured precisely at this stage. This arrangement assures a preset size add-on on warp yarn surface.

The warp sheet is fed to a wet section to facilitate the opening of the yarn surface by improving the dissolvability and reducing yarn hairiness. The fibers and yarns are squashed together in the nipping zone, which posses a negative influence during the drying stage. Generally, two split rods are used to assure a prerequisite movement and slow rotation of size paste. The pre-wetting and washing bath get saturated by accumulation of washed-out components and hence, the liquor must be discharged at preset intervals or rinsed with fresh, hot water to remove the contaminants.

Cylinder drying with overhead drying system

It is very difficult to say that a particular sized yarn drying system is perfect, hence change in the design of drying chamber is continued still, and cylinder drying is one of these technologies. Cylinder drying system was first introduced approximate 160 years ago and still maintains its importance or dominance of the market in the future. The reason for its past success was its extreme economy. The yarn sheet is transported in an assembled form and it has positive effects on the yarn quality, especially during sizing. Modern yarn dryers avoid yarn abrasion, and fibre and yarn hairiness during warp dividing, which occur as the warp yarns rub against each other and during opening of the warp yarns that are clinging together. Invention and application of teflon coating on cylinder surfaces also assists the pressing action of yarn against cylinder surfaces to reduce yarn hairiness and clinging.

Now, cylinder drying technology is equipped with adjustable PC controlled temperature regulating system that works between the pre- and main drying sections to ensure the preset quality, elastic film is produced around the yarns. This widely used drying system is integrated and complemented by a controlled cooling and discharge system to ensure a very controlled sizing action of warp sheet.

The residual moisture content in dried size yarn is measured at the exit of the dryer using three highly sensitive measuring rollers. The actual moisture content is then compared with the target value and, if necessary, a correction action is made by automatically adjusting the production speed.

Warp yarn storage devices are used in the drying zone when producing the warp beams, which enable the sizing machine to continue running even when the beam is being changed. The compensators are designed for handling warp lengths of up to 144 m. They give rise to better and constant sizing values, without over-drying or standing marks, and supply the BM beaming machine. The design of the beaming machine makes it extremely user friendly; for example, the comb is arranged as to permit easy access and facilitate yarn leasing, also makes it easier to deal with any yarn breakages.

BM beaming machine by KARL MAYER

KARL MAYER ROTAL beaming machines work from creel and are able to come up to the highest requirements of the various textile-industry sectors. Highly flexible creel systems for the processing of monofilament yarns, polypropylene film tapes, highly resistant polyester/polyamide yarns for the carpet industry up to glass-fiber yarns can be processed. The two drives on the beaming machine, one for the trio rollers and one for the warp beam, guarantee for all yarn types' regular yarn tensile strengths. Speed between 5 and 400 m/min ensures high efficiency.

You can choose yarn tensile strengths between 5000 and 13000 N (according to the field of application). The user-friendly process control on the beaming machine with its operating area (touch-screen) is another advantage of the machine.

KARL MAYER provided the technology to produce beams up to 1250 mm successfully.

A linear comb traversing facility permits the warp threads to be wound uniformly right up to the edges of weavers beam. A linear comb-traversing facility permits the warp or weaving beams to be wound uniformly right up to the edges. The mechanical, high-speed reversal point of the comb-traversing mechanism prevents the formation of wavy edges during winding.

A three or four roller-control device and a sensitive winding tensioner assure the optimum beam winding pattern and the desired winding length. High-tech processing operations are achieved by making change in a freely programmable beam length, pre-switching device, and a system for reducing the temperature.

KAMCOS®
Operator interface IPC/Touch screen 12, 1", ethernet interface for networking and connection for the relevant data acquisition and teleservice.

Trio rollers
Motorically controlled, for a constant warp-yarn tension

Compact sizing
KARL MAYER, a leading German sizing machine manufacturer has developed a two level (two tier) compact sizing machine to reduce the space requirement of weaving preparatory section as shown in Fig. 4.78. This machine is best suitable to size short length of warp threads. In this machine the let-off section, sizing section, and take-up sections are arranged in a line at floor of the mill. The drying section based on radiation and convection principle by means of infrared drying and hot-air drying respectively is mounted over floor line sections. This compact sizing machine reduces space requirement to almost half of the single-level equivalent sizing machine. This machine has reduced the size requirements, amount of waste yarn, and stop marks.

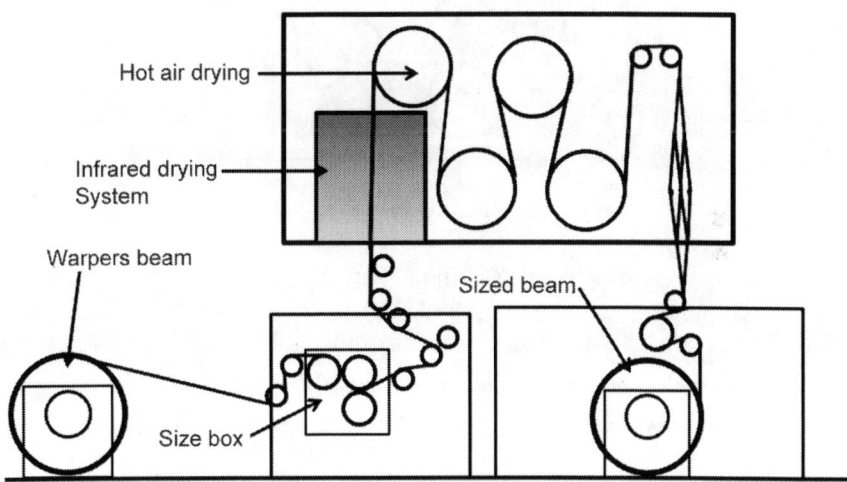

Fig. 4.78 Two level compact sizing by KARL MAYER

On the basis of above facts, it can be safely said that KARL MAYER is continuously developing their products and processes for improving warp yarn performances during fabric manufacturing and fabric manufacturing preparatory within its areas of specialization, "denim" and "sizing" by introducing innovative products and solutions. The major forces of success of KARL MAYER are well focussed product development, combined with forward-looking and innovative processes.

The flexible sizing system

KARL MAYER, Germany developed a flexible sizing concept for cold sizing. In case of heat sensitive warp yarns, size coat has to be applied without application of heat. In this system, sizing paste is applied tangentially. The sizing chemicals in the form of paste are applied by means of a couple of kiss rollers. In this case, indirect heat can be used to dry sized warp threads. This machine can be changed from cold sizing to hot sizing and vice versa as per requirement as shown in Fig. 4.79.

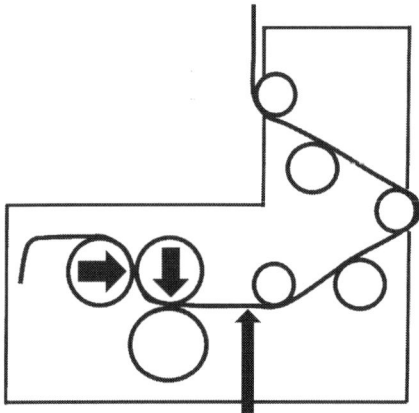

Fig. 4.79 Two-in-one hot and cold size

In compact-size system, a combination of air/infrared drying and by contact, cylinder dryer system is used. Drying zone of compact size is one meter long, contains a gas heater with maximum output of 240 kW, and has a drying performance of 40–50% at a processing speed of 60 m/min.

Development of new size box "SIZE PRO" by KARL MAYER

KARL MAYER has designed a new PRO SIZE box to ensure maximum productivity and flexibility. KARL MAYER's warp preparation machines hold specific yarn tensioners such as, the Rot-O-Tense. KARL MAYER developed a highly innovative compact size box with a trade name of PRO SIZE to apply the size paste at higher efficiency. PRO SIZE is specially designed to process a warp beam of very high thread density even at high size concentrations and viscosities. The PRO SIZE has combined three highly turbulent, flow-intensive application zones, which are produced by pairing three counter-rollers with a common main roller. PRO SIZE size box can work for applying size in the form of liquids, foams, and pastes on warp yarn surface. The

PRO SIZE concept minimizes the amount of processing liquor required, and reduces the volume of effluent produced. This size box has facility to transport the yarn assembly without crossing each other, without twisting, and with setting optimum elongation. The PRO SIZE size box can process 80% of the total number of yarn ends in place of 40% of the total number of yarns on conventional dipping-type size boxes. Two size boxes are advised to be used when the warp sheet width increases up to more than 2000 mm.

PRO SIZE has the potential to increase the maximum warp yarn density to 95%. A single size box PRO SIZE machine can be applied to coat a uniform size film over the entire warp sheet of high warp density warp beams with work width of 2400 mm, 2800 mm or 3200 mm.

Contribution of Toyoda Machines, Japan in the development of sizing process
To fulfill the increasing demand of worldwide customers for both higher productivity and better sizing quality, Toyoda has teamed up with Kawamoto System Corporation, a company with many years of experience in weaving preparatory machine manufacturing. By combining the technological expertise of both companies, Toyoda developed new weaving preparatory machines.

These new technologically advanced weaving preparatory machines accelerated the response from textile industry by introducing digital control via AC vector inverter and other microprocessor based controllers in order to achieve high-quality sizing even at high speed. These machines are operator-friendly, employing the touch-screen function panels. Toyoda machines developed a sizing machine with trade name of "Filamaster Express 610" with all possible developments.

Silent features of Filamaster Express 610
- The tension of the warp yarn sheet is controlled in real time by the AC vector inverter of the let-off and take-up sections. The digital control of machine parts provides precise operation at normal speeds as well as during machine acceleration, deceleration, and stoppage. In this way, a stable yarn tension is possible throughout the process.
- To ensure optimal stretch to the warp yarn sheet, the revolution of squeezing roller and drying cylinder are controlled very precisely which is essential to ensure excellent quality sized yarn. The machine measures the rollers' revolution speed and indicates the stretch in terms of the rollers' revolution ratio.
- With improved circulation of the sizing liquid that overflows from the front and rear of the size box, and a higher quality size-paste filter for the cavity box, bubbles are effectively prevented in the size paste during high speed operation, which can adversely affect the sizing quality. During high speed operation, the sizing paste level is lowered to prevent

splattering.

- Two different circulation routes of sizing liquid remove impurities from the bottom to ensure a constant flow of pure, high-quality sizing liquid to the size boxes for a long run.
- A water-cooled wet splitting rod divides the warp yarn sheet immediately after sizing. The moisture on the rod surface prevents the size paste accumulation on it.
- Hot air is applied in and against the warp yarn sheet's running direction. There are two continuous drying chambers, resulting in much higher drying capacity for the warp yarn sheet.
- The automatic temperature control system using motor valves adjusts the volume of steam and maintains a fixed temperature in the hot air chamber to ensure uniform drying of the warp yarn sheet that remains unaffected from operation speed.
- The drying heat in the hot air chamber is generated by a conventional steam. The performance of weaving preparation process determines loom efficiency, fabric quality, maximum achievable loom speed, and higher productivity in a sizer.

Manufacturers of Filamaster Express 610 made various arrangements to make the machine very user friendly through following arrangements:

Easy operation
- The Filamaster Express 610 is equipped with a user-friendly touch-screen function panel. With the touch of a finger, the screen displays operating conditions such as take-up length, speed, and tension. This facility allows the operator to confirm and adjust the settings quickly.
- The machine stops automatically if a problem occurs during operation and shows the exact location of the problem on the monitor. The temperature indicator records the temperature of six different points from among the size box, hot air chamber, and drying cylinder sections in a graph, giving the operator complete control over the quality of sizing.
- The circulation fan inside the chambers uses an inverter control system, which can automatically set the fan speed to match the operating conditions, to provide consistent, efficient drying, and reduced energy consumption.
- The Filamaster Express 610 includes various safety features for the operator such as covers over the high-pressure cylinders, which become very hot during operation, as well as covers over the drive section and rollers, and an emergency stop button on the main unit.
- The Filamaster Express 610 included high-quality electric heater. With this high drying ability, the temperature can be quickly increased as soon

as the machine starts and during ultra high-speed operation.

- This machine provides per-block temperature control inside the drying cylinders and maintains the quality of the warp sheet very precisely.
- A specially designed equal pressure roller (EPR) squeezes the center and edges of the warp sheet uniformly, preventing irregular sizing of the warp sheet, even during high-pressure squeezing.
- Synchronized control of the machine speed and the squeezing pressure enables the squeezing pressure to adjust according to any changes in operation speed and results stable sizing even during machine acceleration and deceleration.

Prashant Westpoint sizing machine

Prashant Westpoint Pvt. Ltd., Ahmedabad, India has started to manufacture yarn sizing machines for global textile industry. This sizing machine is equipped with VLT Automation drive. The VLT® Automation Drive represents a single drive concept that controls the entire range of operations from ordinary to servo like applications on any machine or production line. The modular open-technology platform, on which the VLT® Automation Drive is built, makes it exceptionally adaptable and programmable. Generally, a sizing machine needs a minimum of 4 to a maximum of 10 drives of range from 1.5 KW to 37 KW. This machine is equipped with VLT® Automation Drive FC 301, an encoder feedback option card that works in closed loop speed feedback and receives speed reference in the form of digital pulses from the dryer.

In this sizing machine, size box, dryer, and drag roll motors are smartly synchronized. Automatic speed controlled winder is used to control the warp tension residual moisture content. The diameter of bare beam is approximately 160 mm. This machine is able to control the tension between the drag roll and the winder.

4.55   Few other sizing machines

4.55.1   Sizing machine for viscose and nylon filament yarn: (Model – Id 0005H)

Id-0005H sizing machine is suitable for nylon and viscose filament yarn. The construction of this machine is very simple and economic. This machine is suitable for process of viscose and nylon filament. This model is beam to beam sizing machine having warper beam stand, sow box having single dip single nip with pneumatic squeezing arrangement, 5 cylinder dryer and head stock with clutch-type DC drive.

4.55.2   Sizing machine for jute industries: Model - Id 0005H

In jute industries, Id 0005H is known as jute dressing machine. For beam width up to 3500 mm and beam diameter 1500 mm, another model IJ 0005H is very common. This machine is available with double beaming attachment. This machine can also be converted to narrow working width with spool creel.

4.55.3   Sizing machine: Model - Id 0005H

Sizing machine Id 0005H is most ideal for sizing more than 6000 ends. This machine is having double sow box and have different version like synchro-2, synchro-4, and parallel sow box. This is suitable for both shuttle and shuttle-less looms.

4.55.4   DW7160 sizing machine

This sizing machine is suitable for warp yarn sizing of cotton, flax, chemical staple fiber, and blending, etc. This machine is manufactured by Dowell Science and Technology Co. Ltd., Hong Kong.

Major features
- Size box
- Single size box, main size box of 240L, aux. box of 180L
- Squeezing rollers covered with rubber, the front roller is center reinforced, suitable for high pressure squeezing
- Size box in steel construction with stainless steel outside casing, good sealing effect, the outside casing is easy to maintain
- Auxiliary size box with automatic liquid level control

Drying
- Direct steam heating with automatic temperature control
- 10 stainless steel drying cylinders (8 cylinders coated with Teflon), 800 mm in diameter
- Stainless steel hood with fan for steam exhaustion
- A water-cooled wet splitting device installed between the size box and pre-dryer
- Warp feeding tension and warp tension between the sizing roller and the dryer (wet zone tension) adjusted by step-less transmission

Creel zone
- Beam creel applicable for 16 beams
- Automatic pneumatic tension control of warp unwinding, maximum let-off tension is 3000 N

- Circulation pump with stainless steel blade and individual motor drive, 20-50 l/min flow rate
- Warp feeding device consisting of feed roller, press roller, and tension measuring roller

Headstock
- Winding drive inverter controlled by AC motor
- Winding done in both directions
- With waxing device, marking device, splitting section, length measuring roller, and tension measuring roller, pressing device, automatic measurement, and control system of moisture regain, etc
- The whole sizing machine controlled by high performance imported industrial PLC, all electrical components confirm to IEC international standards

Machine details

Weaver's beam width	2800 mm
Warper's beam width	1800 mm
Machine speed	1~100 m/min
Crawling speed	0.5 m/min
Squeezing pressure	≤40 kN
Front squeezing	0–40 kN, step-less
Rear squeezing	0–15kN, step-less
Drying capacity	800 kg/h
Weaver's beam diameter	800 mm
Warper's beam diameter	800 mm
Max. winding tension	7 kN
Installed power	
Main drive37 kW	
Auxiliary	13 kW
Weight	35 T

Requirement of warp sizing
The properties which a satisfactory warp size must have may be classified under two general headings:
  (i)   Paste characteristics
  (ii) Film characteristics

(i)   Paste characteristics:
The past characteristics are those properties which govern the behavior of the size during preparation and application. The film characteristics affect its behavior during weaving and subsequently in desizing (Behera at al. 2006).

The size paste characteristics can be described and defined on the basis of the following parameters:

- Ease of preparation
- Uniform viscosity
- Absence of pronounced congealing and skinning properties
- pH near neutral
- Absence of foaming properties
- Stability towards decomposition

(ii) Film characteristics

The size film characteristics can be described and defined on the basis of the following parameters:

- Tensile strength
- Adhesion to the yarn
- Flexibility and folding endurance
- Hardness, especially resistance to abrasion
- Insensitivity to over drying
- Moderate insensitivity to change in relative humidity
- Ready re-solubility regardless of the age of the size film for ease in desizing

4.56   Classification of sizing agents

Classification of sizing agents is shown in Fig. 4.80.

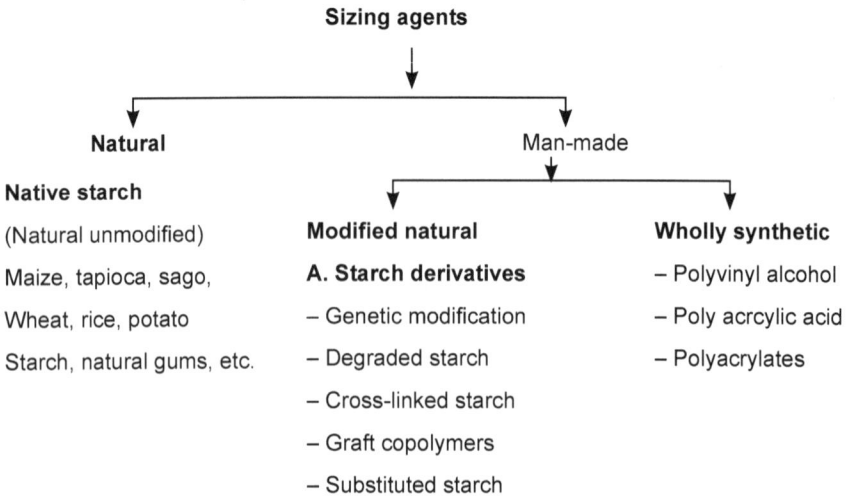

Fig. 4.80 Classification of size material

Unmodified starch

Starch is a reserve carbohydrate obtained from plants where it is generally deposited as minute granules ranging from 1 μm to 100 μm or more in diameter. The shape and size of starch granules vary, depending upon the plant source. These granules are mainly deposited in the seeds, tuber, and roots of the plant. Maize is the major source of starch, though a small amount of starch is also obtained from rice, tapioca, potato, and sago. Chemically, starch is a polymeric carbohydrate consisting of anhydroglucose units (AGU) linked together primarily through α-D-(1–4) glucosidic bonds. It has been established that starch is a heterogeneous material consisting mainly of two major types of polymers, amylose and amylopectin (Figs. 4.5 and 4.6).

Amylose is essentially a linear polymer of 200–2000 anhydroglucose units linked predominantly through α-D-(1–4) glucosidic linkages. Amylopectin on the other hand, is a branched polymer having periodic branching at the carbon-6 position of the amylose chain. The branches contain 20–30 AGU linked to the carbon-6 by α-D-(1–6) glucosidic bonds.

The two fractions occur in varying proportion in the starch from different botanical sources. Both the fractions can be cooked in hot water to a colloidal paste which on cooling forms a rigid, interlocked gel like structure. The linear fraction, amylose, is mainly responsible for the above process called retrogradation.

Starch, in its unmodified form, has been used as a sizing agent for textiles but it needs modification to overcome the limitation. Literature survey shows that the following modifications of starch have been made for sizing of cotton, synthetic, and blended yarns:

- Thin boiling starch
- Cross-linked starch
- Substituted starch
- Grafted starch
- Miscellaneous modifications.

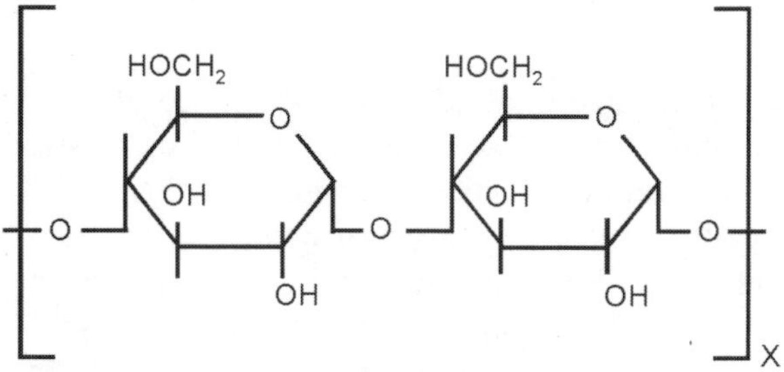

Fig. 4.81 Amylose

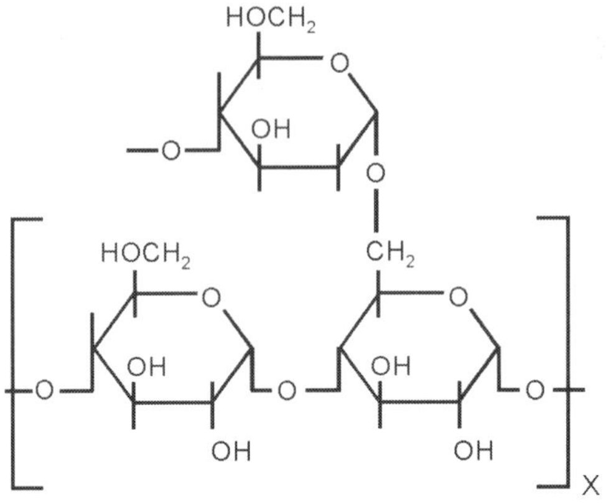

Fig. 4.82 Amylopectin

4.57 Quality and performance assessment of sized yarn

During the journey of grey yarn from yarn spinning system to weaving, the yarn has to withstand various kinds of stresses imposed by different mechanisms (Lord 1973; Fassen et al. 1966; Achwal 1982). The single spun yarns can rarely be woven in grey state (Vincent 1976) as they have higher hairiness and insufficient resistance to abrasion though they withstand normal weaving stresses. The generation of hairiness is the main culprit to the weaving. Iyer (1981) comments that the conventional methods of yarn assessment, count, strength, evenness, and coefficient of variation of strength and count are not sufficient. The yarn defects such as hairiness, gross faults, knots, crackers, etc., are also equally responsible for poor weavability of these yarns. For successful weaving, yarn should not only be strong and uniform but also have maximum possible elongation at break (Annon 1981c).

4.57.1 Evaluation of size material

The properties of the size film, adhesion of the size film to the yarn, and the size penetration decide the weaving performance of the sized yarn. These properties of size material are majorly dependent on the rheological behavior of the sizing paste. From the operational point of view, it is necessary to consider the influence of concentration and temperature on the viscosity of the size to harmonize the adhesion of size material with yarn. The properties of the sizing agents should be evaluated properly for making appropriate technical decisions.

4.57.2   Rheological behavior of size paste

It is well established that the size paste parameters such as concentration and viscosity together determine the add-on of size on any yarn type. Viscosity of size paste under sizing conditions also influences the distribution profile of size within the yarn structure. For homogeneous sizing of warp yarns, the size material must have appropriate and stable viscosity, which is obtained for starch based materials by cooking as described by Langlois et al. (1967). The rheological behavior of starch pastes has been studied under following sections:

- Gelatinization temperature
- Viscosity of cooked paste
- Storage stability and retrogradation

4.57.3   Film properties

The coherent behavior of size film with yarn is the main factor to decide the performance of sized yarn. From research point of view, various sizes are examined to study the properties of size film because under real situation on high speed loom, the size film on yarn surface is subjected to terrific bending and twisting which can lead to the breakage of film and shedding. The size film must have enough elasticity to permit the warp to elongate and return to its original length in case of sizing of man-made fiber yarns as concluded by Hall (1976). Following properties of size film must be effective to evaluate the effectiveness of a particular size:

- Effect of relative humidity on the tensile properties of size film
- Moisture regain of the film
- Mechanical properties of the film

4.57.4   Adhesion

While film-forming properties of the size provide some insight into the abrasion resistance of the sized yarn, one must also consider the adhesion of the size to the yarn/fiber. If the size has good film properties but does not adhere to the yarn/fiber structure, it may be easily peeled-off from the yarn during cyclic bending and abrasive actions. The adhesion of size to yarn depends mainly on the interaction of attractive forces between the size film and the yarn. As the forces associated with hydrogen bonding are much higher than the Vander walls' force, it is reasonable to assume that hydrogen bonding influence is significant in achieving good adhesion as stated by Behera et al. (2008).

4.57.5   Desizing

Sizing is a temporary application used mainly for improving mechanical processing in weaving and the yarn has to be desized for most of the fabrics after weaving. Not only the size ingredients have to be removed easily and efficiently, but it must meet the demands to reduce both water and energy consumption and pollution load of the effluent. It is evident that any change in the technology from the viewpoint of effective desizing must be considered because the degree of size removal depends essentially upon the following factors Hall et al. (2004):

- Size coating/penetration affected by viscosity of the size solution
- The re-solubility of the size film
- Add-on
- The fabric construction
- The desizing process

4.58   Evaluation of sized yarn

Since the purpose of sizing is mainly to improve weavability, changes in the mechanical properties of the yarn on sizing, their relative influence on the weaving performance, the effect of relative humidity on the mechanical properties and weavability of the yarn have been the major fields of investigation of sizing technologists.

4.59   Warp breakage mechanism

A warp yarn is subjected to complex mechanical actions principally of two kinds: wearing action and strain, which cause yarn breakages. Wearing action mainly comes from four sources; rubbing with reed during beat-up, wear due to friction with heald eye, drop wire and shuttle, rubbing of yarn with each other during shed change, friction with lease rods and rolls, etc. Strain comes from two sources; the normal tension on the yarn present on loom and due to shedding, while periodic greater strains occurring during beating. The abrasive action on yarn tries to pull out fibers from yarn body resulting in loosening of yarn structure, which leads to inter-fiber slippage and irrecoverable extension. The continuation of abrasive action further pulls out fibers and a state is reached where the fiber assembly cannot hold together causing a yarn break. It has been observed that in case of unsized grey warp yarns, the yarn breaks mainly due to inter-fiber slippage under various weaving stresses as mentioned earlier. In case of sized yarns, fuzz formation is reduced as surface fibers are partly bound to yarn body. The penetration of size material is responsible to bind fibers inside of the yarn body and thereby delays inter-fiber slippage. Thus, warp breakages are dominated by fiber breakage, and inter-fiber slippage in case of sized ring yarn.

Repeated extension weakens the size film at the faulty points in the yarn. Next the rubbing decreases yarn strength at these weakened points, the effects of rubbing being aggravated by the increased stresses in the size film and fibers. Rubbing plays the principal part during this first stage since it is only rubbing which decreases yarn strength. The decrease in the strength leads to an increase in stress and yarn elongation in this section. The latter accelerates the distortion of yarn and size film by rubbing. The increasing abrasion decreases yarn strength at this point still further with the result that stress, elongation and the effect of repeated extension are increased. As soon as elongation exceeds the endurance limit of the yarn, repeated extension begins to play the dominant part (although both factors continue to act simultaneously), and quickly destroy the yarn structure at that point. The overstressed fibers "tire" and are broken cohesion between then is severed, they separate and are pulled apart and rupture sets in. Broughton concluded that frictional properties of fibers play an important role in yarn failure under tensile loading.

4.60   Weavability

The word weavability means the ability of a set of yarn to perform well on the loom i.e., it denotes the ease of weaving. As the performance of the yarn on the loom depends on some physical properties of the yarn, the testing of sized yarn for related physical properties should give the measure of weavability. However, the unfortunate fact is that there is no direct correlation between the loom performance of the sized yarn and its physical or chemical properties. This does not mean that these two are absolutely unrelated to each other, but a simple quantitative correlation has not been found so far.

Weavability predictor – yarn cohesive strength tester (Hiruta model thread tension rubbing test recording machine)
During cloth manufacturing, warp threads must be resistant to such mechanical strains as tensing, bending, rubbing, shedding, etc., for smooth running of loom without warp breakage. A number of such yarn breakages may lower weaving efficiency and quality of woven cloth.

The conventional yarn testing by tensile strength testers does not provide very reliable data to correlate with weaving efficiency to be known in practical weaving process. Hence, it is very difficult to access and analyze actual weavability of sized yarns by such conventional testers.

To overcome these problems, the Hiruta Riken Co. Ltd. developed a weavability predictor on the basis of a patented technology (patent no. 601116) to simulate the actual loom conditions on this tester. This tester numerically determines yarn cohesive strength.

Weavability predictor developed by Hiruta Riken Co. Ltd. is the only device with patented mechanism (patent no. 601116) to simulate and apply mechanical strains of looms to yarns, and to numerically determine yarn cohesive strength as shown in Fig. 4.83. It is already approved by world scientific community that analysis data obtained by this device correlates to the actual weavability of yarn on loom.

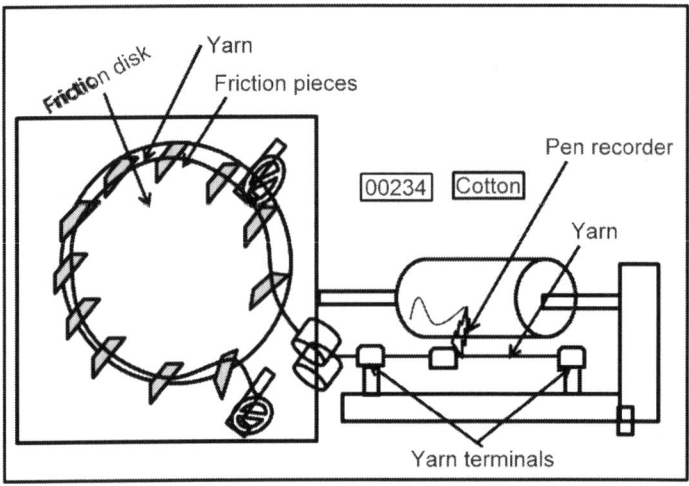

Fig. 4.83 Hiruta Riken weavability predictor

Mechanism

The front face of the device is shown in Fig. 4.83. Yarn to be tested is drawn through the yarn guide. Both the ends of the yarn are held by yarn terminal and clip guide equipped with a specified weight, and the yarn is clipped on recording pen. Turning on the drive switch rotates the friction disk in which the yarn rubs on 11 different-shape friction pieces and receives strains as if it were stressed on a weaving loom. Simultaneously, the recording pen is oscillated on rotating chart paper set on the recording drum. All motions stop when yarn is broken. The number of rotation of friction disk is counted by the counter. Chart drawn by the recording pen may tell the condition of testing yarn particularly when it is elongated.

Weavability test

In order to compare the relative weaving potential of yarns sized by different size material, Reutinger Webtester or other similar instruments are used.

During weaving on a loom a warp, yarn is subjected to various stresses like cyclic extension, axial abrasion, flexing and bending, excluding beat up and yarn entanglements. All these stresses are simulated in Reutinger Webtester. A sheet of 15 parallel threads is considered for Reutinger Webtester at a pre-decided constant tension. The instrument records the number of cycles required to break first 10 threads. The load is decreased by 1/15th of the initial preset load in order to maintain constant tension on each yarn throughout the test for every break.

The average number of weavability cycles of each sample can be found by taking the average of the number of cycles required for breaking first 10 threads on Reutinger Webtester. The instrument also displays the value of yarn elongation resulting due to fatigue and abrasion actions while maintaining constant tension per thread. In case some yarns are elongated and slacked during test, they will be free from abrasion and fatigue. These slack ends do not have any utility on loom and are removed manually at the cycles at which they become slack, and a break is recorded on the instrument. This instrument is able to alter the conditions to perform test to investigate the effect of various yarn tension, intensity of yarn to metal abrasion, yarn cyclic extension, and cyclic speed.

Viscosity test

The viscosity of size paste is an important process parameter in sizing process. Viscosity of size paste must be maintained constant throughout the process ideally, but it varies due to different reasons. In order to minimize the viscosity variation, it must be measured time to time. Various viscometers are available for quick measurement and Brookfield digital viscometer is also one of them. This viscometer has spindles of various shapes and sizes meant for different viscosity levels. It has a rotational speed range of 5–100 rpm. The instrument gives torque applied to spindle for selected rotational speed and multiplying that torque value with constant of spindle, and speed combination is the viscosity of the solution in centipoise.

Cohesion test

Cohesion of size material with fibrous material is an important factor to decide the size performance. Cohesion of size materials is tested by measuring the tensile properties of size film. The prepared size solution is laid on an acrylic sheet uniformly and the film of the paste is made. Acrylic sheet is selected because size material does not stick to the sheet and can easily be stripped off. After drying in normal atmospheric condition, the film is kept inside the desicator for conditioning. The desiccator uses super saturated solution of sodium nitrite which maintains 65% RH. After conditioning for 24 h, the film was cut into 15 cm × 1 cm strips and the thickness of each sample is measured by digital vernier calipers. The specimens are again put into the desiccator for 24 h for conditioning. The conditioned samples are tested on instron tester or any other tensile tester for tensile strength and elongation at break with gauge length of 10 cm and breaking time of 20 ± 3 s. The thickness of the film should be constant and it needs to be taken into consideration. Film strength is the direct measure of cohesive power of size material.

Solubility of size materials
In order to measure the solubility of size ingredient, their films are prepared as mentioned in cohesion testing. The films are cut into strips of equal dimensions. A constant weight is suspended on each strip before subjecting size strips into water, and a constant water level is maintained for all tests as similarly practiced by Behera et al. (2008). The size film starts to swell and looses strength after coming in the contact of water. The time required to break each strip by given weight is recorded. The less time required, higher is the solubility of the film and vice versa.

Retrogradation test
Retrogradation is the change in viscosity with the change in temperature of the size paste. The viscosity of size solution at different temperatures is measured by above-explained method and a graph is plotted which gives the retrogradation property of the size solution. The same method was used for the entire size sample.

Adhesion test
Adhesion potential of size with fibrous surface must be tested. The measurement of the adhesive strength of size material is carried out by applying it on different types of roving, assuming that the original strength of roving is zero or negligible. The rovings are sized similarly by the same method as is used in normal sizing of yarns.

After drying the rovings at room temperature, rovings are conditioned in desiccators for 24 h at 65% RH. Strength of roving is tested on any tensile tester like instron with gauge length preferably 20 cm and breaking time of 20 ± 3s. The strength of sized roving is the representation of the strength of size or the adhesive potential of size material.

Test for moisture regain of size films
In order to calculate the precise moisture regain, the size films are conditioned in desiccators for 24 h at 65% RH. After weighing of size films carefully, they are kept inside hot-air oven for 30 min at $105 \pm 3°C$. The oven dry weight of size film is measured. Moisture regain of size film is calculated by using the following formula:

$$\text{Moisture regain (\%)} = \frac{W - D}{D} \times 100$$

Where W is the weight of size film at 65% RH, and D is the oven dry weight of size film.

4.61   Mechanical and physical properties of yarn

The performance of sized yarn has to be measured in terms of their mechanical and physical properties. The tensile properties are breaking strength, breaking extension, work of rupture, and initial modulus while the physical attributes are hairiness of the yarn, bending/flexural rigidity, abrasion resistance, sonic modulus, and effect of relative humidity on the sized yarn.

Desizing efficiency test
Desizing efficiency is found in two ways, conventional and TEGEWA method. To perform this test:
Reagent: potassium iodide (10 g of KI (100%) in 100 ml water, add 0.6358 g of iodine, (100%)) stir and shake till iodine is completely dissolved. Fill up to 800 ml with water then complete it to 1000 ml with ethanol. Shelf life is approx 6 months only.

Method
  1. Spot drop-wise solution onto fabric
  2. Rub in gently
  3. Assess change of color

Note: the test must be carried on fabric that has been cooled down to room temperature residual alkalinity has to be neutralies prior to the test
Assessment
  (a) Grey fabric
       No change of color = no starch size present
       Pale blue to bluish = presence of starch size or blend
       Violet = of starch size with synthetic size
  (b) Desized fabric
       Pale blue to bluish violet = violet scale TEGEWA. This indicates residual Starch content.
       Moisture regain of sized yarn (after drying process) for smoothness/softness.

Conventional method:
In this method we first take the weight of the sized fabric, let it be W1. Then desize the fabric, dry it and take the weight, let it be W2. After that, the fabric

is treated with 3gpl (35%) HCl at 700°C for 30 min. Dry and take the weight of the fabric. Let it be W3.

Total size = W1 – W3.

Residual size = W2 – W3.

Desizing efficiency = (Total size – Residual size)/ Total size × 100.

Residual moisture content in sized yarn

During drying of sized yarns on sizing machine, neither over drying nor under drying is advised. An optimum drying gives better performance on loom. The residual moisture content in sized yarn should be kept around following figures:

Type of yarn	Residual moisture content
100% cotton	6.5–7.0%
100% rayon	9.0–9.5%
100% polyester	1.0–1.5%
65/35 polyester/cotton	3.0–3.5%

Size coat determination

Size paste is applied in order to enhance the weaving performance of warp yarns. Optimization of size add-on is essential to save the material and power both. Optimization is performed by appropriate testing of size add-on size yarns. Various techniques are developed by different scientist to evaluate the size add-on on warp threads.

Desizing

Initially, desizing was started to determine the size add-on on warp yarns accurately. In this process, the specimen is rinsed and dried out perfectly. This is an off-line testing method, and is carried out in a laboratory; on line control of size add add-on during sizing is not possible by this method. By desizing method, the yarn is cooked for some time in a defined bath. The remarkable disadvantage of this technique is that a part of the fibers and other additions is removed, so that a higher value of size coat is determined than the real one; as it is calculated by equation

$$S_{1,1}, S_{1,2}, S_{1,3} = \frac{G_{ds} - G_{dr}}{G_{dr}} \times 100 \,\%$$

where $S_{1,1}, S_{1,2}, S_{1,3}$ size coat in percent is obtained by desizing (using chloride acid, soft water or distilled water)

G_{ds} mass of dry sized yarn in grams (g)

G_{dr} mass of dry desized yarn obtained after cooking and rinsing in grams (g)

Drying

Drying technique can also be applied to calculate size add-on on warp yarn.

This technique is based on the measurement before immersing specimen into size paste and sized warp specimen at the exit point of size box for same length. Both samples are dried perfectly and weighed, and then the following equation can be used to calculate the size add-on %.

$$S_2 = \frac{G_{ds} - G_d}{G_d} \times 100\%$$

S_2 – size coat obtained by drying the samples %
G_{ds} – mass of the dry sized yarn in grams (g)
G_d – mass of the dry unsized yarn in grams (g)

In an alternate method, warp size add-on % can be calculated by drying and weighing unsized samples (before immersing warp into the size) and sized warp samples at the exit of the size box (both of the same length). Finally, the difference of their weight is used to calculate the size coat by the following equation

$$S_2(\%) = \frac{G_{ds} - G_d}{G_d} \times 100$$

S_2 – size coat obtained by drying the samples, %;
G_{ds} – mass of the dry sized yarn, in grams (g)
G_d – mass of the dry unsized yarn, in grams (g)

Above discussed methods are sluggish; hence by the application of infrared devices for the drying of samples, quick result calculation is possible. According to Trauter (1982), this method is very precise and quick, accepted by the majority of scientific community worldwide. In this method, the size add-on calculation can be performed by taking a wet sized warp sample from the sizing machine just before drying, weighing it before and after drying, and determining the size concentration by following equation (Kovačević et al. 2002):

$$S(\%) = \frac{\dfrac{W \times K^*}{1 - K^*}}{(G_{ws} - w) - \dfrac{W \times K^*}{1 - K}} \times 100$$

Where,
S – size coat, %
G_{ws} – mass of the wet sized sample before drying, in grams (g)
G_{ds} – mass of the dry sized sample, in grams (g)
$W(g) = G_{ws} - G_{ds}$
K* – effective concentration
K_f – correction factor (the use of correction factor is optional, depending

on size agents used)

K – concentration of size coat, %. This method is considered as one of the best for determining the size coat, but it is not suitable in practice, so it is avoided.

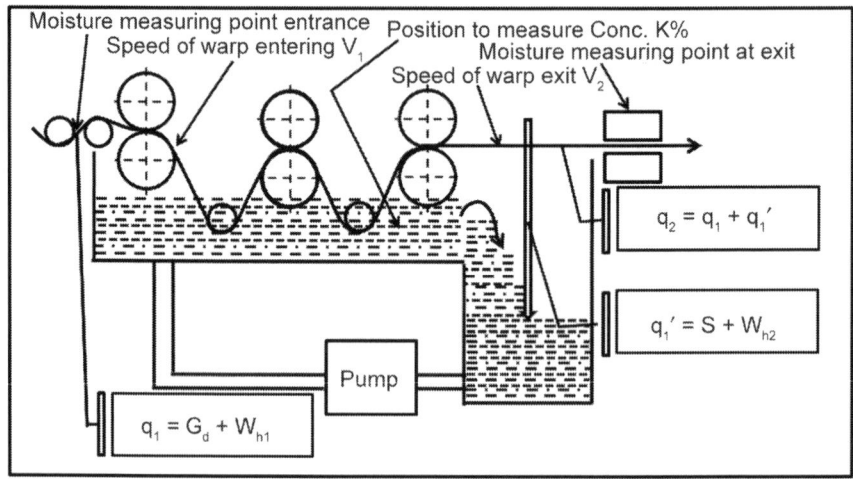

Fig. 4.84 Shirley size box

Shirley size box (Fig. 4.84) is very effective and precise to maintain the uniform size coat but after definite time intervals, dozing of size paste is essential to maintain the constant size level and uniform size pick-up. Water is added to maintain a constant level of the size in the first (pre-wetting) size box. Size concentration is maintained by correcting the dosage of stock size, and the squeezing force varies depending on the machine speed. This design of size box has not been accepted in practice despite its advantages. The reason may be the inadequate size preparation in the sizing machine (Trauter et al. 1982; Kovačević et al. 2002).

The Sucker Muller Company has credited to develop Visacool system of measurement and regulation of size add-on in the sizing process where the size coat was automatically and continuously maintained through the whole sizing.

4.62   Infrared reflection (IR) method

Amirbayat et al. (1995) developed a system to measure the residual moisture content in sized warp yarn continuously by using a contact-free moisture measuring system at the exit of the size box that uses the achievements of both electronics and optics. These instruments can be divided into the devices measuring warp moisture by means of â-radiation absorption. FH 46 device manufactured by Frieseke and Hoepfner GmbH, infra-red IR reflection HMFIRP device manufactured by Mahlo GmbH + Co. KG, and microwave absorption AF 310 device were manufactured by Pleva GmbH. This system is able to control size add-on % by altering the force on the last pair of the squeezing rollers. During moisture measurement and calculation, calibrations are carried out according to the raw material composition of the warp, density, yarn count, size agents, and other parameters affecting measuring signals (Cote 1991; Eickmeier et al. 1991; Stegmaier 1990). This device works by using the absorption of β-radiation of radioactive isotopes, e.g. ^{85}Kr for measuring range 50–1000 g/m² ($\pm$ 1 g/m²) or at ^{90}Sr for the range of 250–2500 g/m² ($\pm$ g/m²). This method can measure 10–5000 g/m². Even after having all these merits, this method has not been accepted commercially due to its high cost and its sensitivity during production conditions (Wegener 1972 and Wildmann 1981).

Infrared (IR) device measures the moisture by using two different wavelengths; one sensitive to moisture changes, and the other having almost no sensitivity to warp moisture, or with a minimal reaction. The monochromatic light is used to measure the moisture content of warp threads. The intensity of reflection with sensitive wavelength of monochromic light depends on the warp moisture content. Typically, the wavelengths 1450 nm and 1930 nm produce a very high sensitivity by changing moisture in the infra-red range. As the moisture content varies by a very small fraction, the intensity of reflection with a different water portion, e.g. in the warp, will lay in the range of wave lengths 1200–1300 nm and in the range 1700–1850 nm, especially by using older measuring instruments that are not sufficiently sensitive to signals.

$$S_3 (\%) = W_{h3} \frac{K}{100 - K}$$

Where,

S_3 – Size add-on %

W_{h3} – Outlet moisture of the warp, %

K – Concentration of the size, %

4.63   Microwave absorption method

This is carried out by contact-free measurement of the solution or dispersal agents by microwave absorption through the permanent electric-dipole moment which causes dielectric losses. Over a time interval, the permanent dipole moment will react positively or negatively to the charge centre. The stronger the bond of water molecules is, the lower the dipole moments and microwave damping. The warp with a higher percentage of water produces more cases of water bonding; e.g., hydrogen bonds between close water molecules, hydrogen bonds between water molecules, and –OH cellulose groups or physical adsorption forces in the surface layer between water and fibers. Accurate measurement of absolute warp moisture is based on this interpretation, because calibrating water is bonded differently with various size agents. This is one of the simplest methods of determining the size coat, and its maintenance during sizing is mathematically based. It constantly measures warp moisture at the exit of the size box, and the size concentration in the size box. Many research projects have dealt with this method, and it has been proved to be the most suitable and effective one. All the authors used the same mathematical procedures (equation 4), but various measuring units. Constant determining of the size coat is done by taking into consideration the following three parameters: inlet moisture, outlet moisture, and concentration, and by equation (5). This final equation was obtained by analysis of the substance content:

$$S_4 = \frac{W_{h3} - W_{h1}}{\dfrac{100}{K(100\%} - \dfrac{W_{h3}}{100} - 1}\%$$

Where,

S_4 – Size add-on, %

W_{h3} – Outlet moisture of the warp, %;

W_{h1} – Inlet moisture of the warp, %;

K – Concentration of the size, %.

4.64 Process control in sizing and sizing defects

4.64.1 Moisture control in sized beam

A moisture control of 8–10% should be maintained in the sized cotton yarns. Excessive drying makes the size film brittle and harsh; but very high moisture content makes the size film soft and the yarns become sticky.

4.64.2 Yarn stretching during sizing

During sizing, the yarns are subjected to various types of stresses which result in a permanent elongation in the yarn. It leads to a decrease in residual extensibility or elongation at break of the sized yarn, which leads to more breakage at the loom shed. There are some specific zones where stretch control of sized yarns is essential.

1. Creel zone

Creel zone starts from last warper beam and ends at dry nip. The creel stretch on existing type of conventional sizing machines is generally controlled manually but latest sizing machines are capable to control creel stretch by adjusting the speed of first pair of roller where warp sheet nips using programmed speed control. The yarn tension in creel section increases gradually with reduction in warper's beam diameter. To overcome this problem, the loading on warper's beam must be adjusted in controlled way as sizing progresses and warper's beam progresses.

2. Wet zone

Wet zone starts from dry nip and ends at first drying cylinder. The control of yarn stretching is performed by making the all positive drives to various rollers and cylinders of wet zone by using dedicated servo motors.

3. Drying zone

Drying zone starts from first drying cylinder and ends at last drying cylinder. The stretch control in drying zone can take place by synchronizing the PIV gear during the cylinders and with that, driving the finishing squeeze roller in wet zone.

4. Splitting zone

Splitting zone starts from last drying cylinder and ends at drag roll. The stretch control in splitting zone can be controlled by providing a dedicated synchronized servo motor to control the speed of drag roll precisely.

5. Winding zone

Winding zone starts from drag roll and ends at weavers beam. At constant RPM, winding speed increases with increase in beam diameter. The stretch control in winding zone can be achieved by controlling the speed of weavers beam very precisely with the help of PID controllers attached with PIV gearing or dedicated servo motor for weaver's beam.

Quality of sized beam

A high quality weaver's beam should unwind smoothly during fabric manufacturing. These are some very commonly occurring faults of sized beam.

Density variation in weaver's beam

Density of weaver's beam should be homogeneous throughout the winding from starting to the end point. Loosely wound weaver's beam creates lots of problem during fabric manufacturing, particularly loose ends. However, highly dense weavers beam restricts the smooth release of warp sheet from weavers beam. Density of weaver's beam is mainly influenced by two factors.

- Performance of friction clutch or the DC drive
- Effectiveness of the beam pressing motion

Broken ends, missing ends, crossed ends, and sticky ends
All these faults are very common in weaver's beam and the major causes of these faults are
- *Formation of lappers:* A lapper is an accumulation of layers of yarn on the weaver's beam.
- *Invisible breaks during sizing:* The end breaks which take place during sizing that does not form lappers are called invisible breaks.

Both lappers and invisible ends become the source of missing and broken ends in the sized beam.

Sticky ends
Sticky ends come in picture when broken ends from one warper's beam migrate to the yarn of another warper's beam.

Formation of ridges on beam
Ridges are formed on the beam when the ends taken in one dent of the comb are not spread-out properly. To overcome these defects, effective eccentric dancing rollers must be adjusted at the headstock of sizing machine.

Crossed ends
Crossed ends are formed during weaving whenever the leading end is not available in the appropriate place on the beam, and therefore, the weaver has to knot the trailing end to an end that is far away. This happens because in some cases, the leading end of an invisible break migrates to a distant place.

Most of the above mentioned faults occur due to the happening of lappers and invisible ends. Hence, serious and consistent efforts are required to control these defects.

Factors affecting invisible ends:
- High thread 8stretch during yarn sizing
- Frequently repeating weak place in yarn

Factors affecting lappers:
- Condition of beam flanges
- Efficiency of warp stop motion
- End breakage rate at warping

Factors affecting defective selvedges:
Generally, two types of defective selvedges are more serious hindrance in unwinding during fabric manufacturing in comparison of body ends of the beam. These selvedges are

- Bulging selvedges
- Sunken selvedges

These defects can be effectively controlled by

- Correctly setting the expandable comb at the headstock
- Assuring that beam flanges are perfectly usable
- Using the judicious size of beam pressing roller so that it reaches both the beam flanges

Some research related to sizing

Kovacevic et al. (2008) analyzed the effect of sizing on multicolored warps and inferred in following manner

- Differently colored yarns showed a different amount of the size pick-up after sizing in homogeneous conditions.
- The difference in properties prior to sizing affected the properties of the sized yarn, as well as the size pick-up.
- After dyeing, the bleached yarn showed the greatest changes in physical mechanical-properties.
- Breaking force and elongation at break of yarn after dyeing differed in colors, although the raw yarn was the same.
- The difference between the white yarns of different warps under the same sizing conditions did not show either in the size pick-up or abrasion resistance.
- These results lead to the conclusion that in yarn dyeing playing major role in deciding size pick-up and that may attribute to change in yarn properties.
- Some dyes close threads in such a way that the size does not penetrate thread interspaces, resulting in a lower size pick-up.
- The bleached yarn always had a lower size pick-up compared with other colors. In comparison of other colors, it may be concluded that darker colors have a lower size pick-up.
- Blue color had a lower size pick-up than yellow and black had a lower size pick-up than orange and inferior physical-mechanical yarn properties too.

Schwarz et al. (2011) studied the degree of impact in mechanical and deformation properties of single cotton yarn after sizing depending on their fineness, size concentration and the structure created by the type of yarn spinning. PVA sizing agent was used on both ring and rotor yarns using the same recipe with two different size concentrations 7.5% and 5.0%. Modified method for cyclic testing was also used along with some traditional mechanical tests to determine areas of elastic, viscoelastic, and plastic yarn deformations. Conclusion of the study revealed that no remarkable differences in yarn properties after sizing with lower and higher size concentrations, was observed.

These findings lead to the conclusion that it is a greater advantage to use a lower concentration, in order to achieve better economic and environmental effects. Ring spun yarns exhibit major improvements, which are positive signs because improvements in mechanical properties of ring-spun yarns are more necessary.

Contrary to this, in case of rotor-spun yarn the improvements are relatively low, due to its unique structure. It is proved by cyclic loading of yarn that rotor yarn structure is the key factor to register a difference in changes of deformation areas of unsized and sized yarns, primarily in the way of reducing the elastic area of deformation.

4.65   Classimat faults

As per Doleckie (1974), the contribution of thick places in yarn breakages during fabric manufacturing is significant and on an average depending on yarn quality, 30–70% warp breaks are due to thick places or slubs either on the same thread (20–40%) or on adjacent threads (30–50%). The rest are due to other reasons such as tension peaks, weak or thin places, and knots in the yarn, etc. This study highlights the significance of identifying and removing warp-break prone thick places during winding.

The tensile behavior and abrasion resistance of yarns mainly decide the fabric manufacturing potential of any yarn. Using this concept and approach, Aggrawal et al. (1987) studied the impact of Classimat faults on weavability of yarns on the basis of magnitude of yarn faults (length and cross section) and sizing. Their study concluded with the inferences that the yarn having C3, C4, and all D Classimat faults, even after sizing, have lower strength, elongation, and abrasion resistance. These faults recorded a significant reduction in strength, elongation, and abrasion resistance. During tensile testing, most of the yarns broke at or in the vicinity of a fault. The increased fault length and cross section enhance the frequency yarn breakages. The strength and elongation of grey normal yarn is not affected by the presence of A3 and A4 faults in yarn. The realization of strength and elongation is minimum in the case of D3 and D4 faults. With sizing, except for D3 and D4 faults, the realization factor for the strength and elongation of the faults decreases. The sizing process improves the strength of yarns having frequent D3 and D4 faults. The frequency of yams that break at or in the vicinity of a fault is increased with increasing fault length and fault cross sectional area.

4.66   Yarn numbering system

Direct yarn numbering system

Term	Definition
Tex	Mass of yarn in gram per 1000 meter of yarn length: $$Tex = \frac{mass\ in\ grams}{1000\ meter\ length}$$
Decitex (dtex)	Mass of yarn in gram per 10,000 meter of yarn length: $$dtex = \frac{mass\ in\ grams}{10000\ meter\ length}\ or\ \frac{1}{10}tex,\ \text{decitex is the direct yarn}$$ numbering system recognized by all international bodies for filament and spun yarns in manmade fiber industry
Kilotex (ktex)	Mass of yarn in kg per 1000 meter of yarn length: $$ktex = \frac{mass\ in\ kg}{1000\ meter\ length}\ or\ 1000\ tex,\ \text{ktex is the direct yarn}$$ numbering system recognized by all international bodies for spinning tow and other similar products in manmade fiber industry.
Denier (den)	Mass of yarn in gram per 9000 meter of yarn length: $$denier = \frac{mass\ in\ grams}{9000\ meter\ length}$$

Indirect yarn numbering system

English cotton count (Ne)	Number of 840 yard strands (hank) per 1 English pound (lbs) of mass $$Ne = \frac{Number\ of\ strands\ of\ 840\ yards\ length}{mass\ in\ 1\ pound\ (lbs)}$$
Metric yarn count (Nm)	Number of 1000 meter strands per 1 Kilogram (kg) of mass $$Nm = \frac{Number\ of\ strands\ of\ 1000\ yards\ length}{mass\ in\ 1\ kg}$$
Wool worsted count (NeK)	Number of 560 yards strands per 1 English pound (lbs) of mass $$NeK = \frac{Number\ of\ strands\ of\ 560\ yards\ length}{mass\ in\ 1\ pound\ (lbs)}$$
Wool woolen count (NeS)	Number of 256 yards strands per 1 English pound (lbs) of mass $$NeS = \frac{Number\ of\ strands\ of\ 256\ yards\ length}{mass\ in\ 1\ pound\ (lbs)}$$
Linen count (NeL)	Number of 300 yards strands per 1 English pound (lbs) of mass $$NeL = \frac{Number\ of\ strands\ of\ 300\ yards\ length}{mass\ in\ 1\ pound\ (lbs)}$$

Conversion of different yarn numbering system

Convert into → known ↓	Tex	Decitex (dtex)	Denier (den)	Metric count (Nm)	English Count (Ne)
tex		10 × tex	9 × tex	$\dfrac{1000}{tex}$	$\dfrac{591}{tex}$
decitex (dtex)	$\dfrac{dtex}{10}$		0.9 × dtex	$\dfrac{10000}{dtex}$	$\dfrac{10000}{dtex}$
denier (den)	$\dfrac{den}{9}$	$\dfrac{den}{0.9}$		$\dfrac{9000}{den}$	$\dfrac{5314}{den}$
Metric count (Nm)	$\dfrac{1000}{Nm}$	$\dfrac{10000}{Nm}$	$\dfrac{9000}{Nm}$		0.59 × Nm
English count (Ne)	$\dfrac{591}{Ne}$	$\dfrac{5910}{Ne}$	$\dfrac{5314}{Ne}$	Ne × 1.69	

A typical example for conversion of yarn numbering: A typical 20 tex is converted into various yarn numbering systems.

Selected unit		Unit definition
Tex	20	grams/kilometre
Denier	180	grams/9000 m (or 9 km)
Linen (dry spun), hemp, jute	5.81	lb/14,400 yd (spyndle)
Woolen (Aberdeen)	5.81	lb/14,400 yd
Woolen (US grain)	5.64	grains/20 yd
Asbestos (American)	248	100 yd (cut)/lb
Asbestos (English)	496	50 yd/lb
Cotton bump yarn	1550	yd/oz
Cotton (English)	29.5	840 yd (hank)/lb
Glass (UK and USA)	248	100 yd/lb
Linen (wet or dry spun)	82.7	300 yd(lea)/lb
Metric	50	km/kg
Spun silk	29.5	840 yd/lb
Woolen (American cut)	82.7	300 yd/lb
Woolen (American run)	15.5	100 yd/oz
Woolen (Dewsbury)	1550	yd/oz

Contd...

Selected unit		Unit definition
Woolen (Galashiels)	124	300 yd (cut)/24 oz
Woolen (Hawick)	134	300 yd (cut)/26 oz
Woolen (Irish)	388	yd/0.25 oz
Woolen (West of England)	77.5	320 yd (snap)/lb
Woolen (Yorkshire)	96.9	256 yd (skein)/lb
Woolen (Yorkshire)	96.9	yd/dram
Worsted	44.3	560 yd (hank)/lb
Tex	20	g/km

lb = pound(s)   yds = yard(s)   oz = ounce(s)

References

- Annon, Textile Manufacturers (4) 1977a, p. 41.

- Aggarwal, S. K., and Subramaniam, T. A. (1983), ATIRA Tech digest, **17**(1), p. 18.

- Achwal, W. B., "Refresh your weaving."

- Annon, Text. Mafr. Oct. (4) 1977b, p. 38.

- Aggrawal, S. K., Hari, P. K., Subramanian, T. A. (1987). "Evaluation of Classimat Faults for Their Performance in Weaving" , *Textile Research Journal,* **57**(12), p. 735.

- Annon, Text. World. (1981), **111**(5), p.42.

- Azarchab, M., and Murrowlees, H. (1982), Text. Prax. Int., **37**(11),p. 191.

- Abraham, E. T., and Simi, C. K. (2007),"Hydrophobic Grafted and Cross-linked Starch Nano-particles for Drug Delivery", Bioprocess & Biosystem Eng. 30(3), pp. 173–180.

- Annon, Textile Manufacturers (1977a), (4), p. 41.

- Aggarwal, S. K., and Subramaniam, T. A. (1983), ATIRA Tech digest **17**(1), p. 18.

- Annon, Textile Manufacturers (1977b), (4), p. 38.

- Amirbayat, J., and Bozzalta, G. (1995), *Textile Res. J.* **65**(12), pp. 760–764.

- Annon, Textile World (1981c), **111**(5), p. 42.

- Azarchab, M., and Murrowlees, H. (1982), Text. Prax. Int., **37**(11), p. 191.

- Brancroft, W. D. (1934), "Applied Colloid Chemistry" McGraw, New York.

- Behera, B. K., Mishra, R., and Nakum, S. (2008), Comparative Analysis of Size Materials and Their Weaving Potential, *Industrial Journal of Fiber & Textile*. Res. **23**, pp. 132–138.

- Bowman, L. E., Caley, C. G., Hallen, R. T., and Fulton, J. L. (1996), Sizing and Desizing Polyester/Cotton Blend Yarns Using Liquid Carbon-dioxide, *Textile Res. J.,* **66**(12),p. 795.

- Bowman, L. E., Reade, N. H., and Hallen, R. T. (1998), Advances in Carbon Dioxide Based Sizing and Desizing, *Textile Res. J.*, **68**(10),p. 732.

- Benninger: www.benningergroup.com

- Barella, A. (1957), Yarn Hairiness: The Influence of Twist, *J. Textile Inst.* **48**, p. 268.

- Behera, B. K., and Mishra, R. (2006), *J of Text. Eng.* **52**(5), pp. 179–187.

- Coté, B. et al. (1991), *Textile Res. J.* **61** (12), pp. 724–728.

- Cui, S. W. (2005), *Food carbohydrates: chemistry, physical properties, and applications,* Taylor & Francis/CRC Press: Boca Raton, FL, USA.

- Cheng, K. P. S., and Lam, H. L. I. (2005), "The Appearance of Pneumatic Spliced Short Staple Ring Spun Yarns" *RJTA,* **5**(1).

- Cornuéjols D (2009). Biological crystals: at the interface between physics, chemistry and biology. *Science in School* **11**: 70–76.

- Das, A., Ishtiaque, S. M., and Parida, J. R. (2005), "Effect of Fiber Friction, Yarn Twist, and Splicing Air Pressure on Yarn Splicing Performance" *Fibers and Polymers,* **6**(1), pp. 72–78.

- Doleckie, S. K., (1974) "The Causes of Warp Breaks in Weaving of Spun Yarns", *J. Textile Inst.* **65**, T68–T74.

- Exbrayat, P. E. (1992), A Future Way to Sizing Taking Account of New Spinning and Weaving Technologies, *Melliand Textilber,***73**(1), pp. 28–32.

- Eickmeier, A. (1991), Textil Prax. Int., **46** (5), pp. 458–461.

- Fiebig, D., and Soltau, D. (1995), Some Controlling Factors of Desizing Cotton Yarn, *Textilveredlung,* **30**(5/6), pp. 117–121.

- Fassen, N. J., and Harten, K. V. (1966), *J of Text. Inst.* **57**(7), T 269.

- Farrow, and Hall (1980), Theory and Practice of Textile Slashing, Alabama Textile Operating Executive's Committee, Auburn University, Auburn, Alabama.

- Goswami, B. C., Anandjiwala, R. D., and Hall, D. M. (2004), Textile Sizing, Marcel Dekker, Inc., New York, Basel, ISBN: 0-8247-5053-5.

- Gallant, D. J. Bouchet, B. (1986), Ultrastructure of Maize Starch Granules: A Review, Food Microstr, **5**, pp. 141–155.

- Gotlieb, K. F. (2005), "Starch Derivatization: Fascinating and Unique Industrial Opportunities," Wageningen Academic Publishers, The Netherlands.

- Harper C A (2002), Handbook of plastics, elastomers, and composites, McGraw-hill Publication, Fourth Edition Ed, New York, Chicago, ISBN: 9780071384766

- Houghton, E. F. (1982), "Cotton Warp Sizing Handbook" A publication of the Houghton Textile Research Staff, Canada.

- Habibi, Y., and Lucia, L. A. (2012), "Polysaccharide Building Blocks: A Sustainable Approach to the Development of Renewable Biomaterials," John Wiley & Sons.

- Hall, D. M. (1976), Some Factors Affecting the Application of Size to Textiles. *Text. Inst. Ind.,* **14**, pp. 197–201.

- Hassen, M. B., Jaouachi, B., Sahnoun, M., and Sakli, F. (2007), "Mechanical Properties and Appearance of Wet-Spliced Cotton/Elastane Yarns" *Journal of the Textile Institute,* **98**(1), pp. 119–123.

- Hassen, M. B., and Sakli, F. (2005), "Experimental Study of Mechanical Splicing" *Journal of the Textile Institute,* **96**(3), pp. 111–116.

- Hall D M, Goswami, B. C., and Anandjiwala, R. D. (2004), Textile Sizing, Third Chapter "The Chemistry of Sizing Compounds" CRC Press, Print ISBN 978-0-8247-5053-4.

- Houghton, E. F. (1982), "Cotton Warp Sizing Handbook" A publication of the Houghton Textile Research Staff, Canada.

- Iyer, S. S. (1981), Textile Asia, 12(5), p. 46.

- Jenkins, P. J. and Donald, A. M. (1995) "The influence of amylose on starch granule structure". *Int. J. Biol. Macromol.,* 17, 315–321.

- Katovic, D., Vukusic, S. B., Grgac, S. F., Kovacevic, S., and Schwarz, I. (2008), "The Effect of Microwave Drying on Warp Sizing" *Textile Research Journal,* **78**, p. 353.

- Koch, H., and Roper, H. (1988). New Industrial Products from Starch. *Staerke,* **40**, pp. 121–131.

- Kovacević, S. (2004), Impact of Sizing on Physical-mechanical Properties of Yarn, *Fibers Textiles East. Eur.,* **48**(4), pp. 32–36.

- Kovačević, S., Grancarić, A. M., and Stipančić, M. (2002), Determination of the Size Coat Fib. &Text. In East. Eur., pp. 63–67.

- Kroezen, A. B. J., and Wassink, J. G. (1986), "Foam Generation in Rotor-Stator Mixers" J. of Soc. of Dyers and Col. Vol. 102(12), pp. 397–402.

- Kovacevic, S., Schwarz, I., and Brnada, S. (2008), Analysis of Size Pick-up and Mechanical and Surface Properties of Multicolored Warps, *Text. Res. J ,* **78**(2), pp. 158–167.

- Kovačević S, Grancarić A M and Stipančić M Determination of the Size Coat Fib &Text. In East. Eur. (July/Sep.) 2002, 63–67.

- Lord, P. R. (1973), "Weaving: Conversion of Yarn to Fabric," Marrow Pub. Co. Ltd. England, p. 99.

- Langlois, D. P., Wagoner, J. A., Whistler, R. L., and Paschall, E. F. (1967), Production and Uses of Amylose, *Starch Chemistry and Technology.* Eds.; Academic Press: New York, Vol. II, p. 456.

- Lang, J., Zhu, S., and Pan, N. (2004), "Changing Yarn Hairiness During Winding; Analyzing the Trailing Fiber Ends" *Textile Research Journal,* **74**(10), p. 905.

- Langston, J. H. (1952), Sizing of Spun Nylon Yarns with Various Agents, *Textile Res. J,* Feb., pp. 111–116.

- Lord-1, P. R., and Mohamed, M. H. (1982), "Weaving: Conversion of Yarn to Fabric," Merrow Technical Library, Durham, England, p.108.

- Langer, J. (1972), Mell Text., **66**(12), pp. 20–22.

- Langston, J. H. (1952), Sizing of Spun Nylon Yarns with Various Agents, *Textile Res. J,* pp. 111–116.

- Modi, J. J. (1983), "Recent Advances in Low-cost Sizing," Textile Industry, Vol.–147, 4, p. 58–61.

- Morton, W. E., and Hearle, J. W. S. (1962), "Physical Properties of Fibers" Manchester and London, The Textile Institute, Butterworth.

- Mazíková, V., Sroková, I., and Ebringerová, A. (2009), Chemical Papers **63**(1)pp. 71–76.

- Mostafa, Kh. M. (1997), "Synthesis of Poly (acrylamide)-starch", *Polymer Degradation and Stability,* **55**, pp. 125–130.

- Namboodri, C. G. (1986), Foam Sizing of Cotton and Blend Yarns: Slashing Trials *Text. Res. J,* **56**(2), pp. 87–92.

- Ok, H., Carr, W. W., and Park, H. (2007), "Single-end Sizing of Yarn Using a Slot Applicator" *Text. Res J,* **77**, pp. 686–695.

- Porat, I., Greenwood, K., and Banukda, Z. (1999), "Computer Simulation of Cone Winding Mechanisms with Constant Rate of Yarn Delivery systems", *Textile Research Journal,* **69**(2), pp. 113–120.

- Paul, A. (1982), Length Measurement in the Winding Department, Int. Textile Bull Weaving, p. 2.

- Pleva, R., and Rieger, W. (1992), Measurement and Optimization of Size Pick-up, Textile Praxis Int., Vol. **47** (3), pp. 230–232.

- Platt sizing (1980), "Solvent Sizer Underscores Importance of Energy," Textile World, **130**(10), pp. 100–102.

- Perkins, W. S., and Walker, R. P. (1982), "Foam sizing" *Text. Res. J,* 52, pp. 547–554.

- Perkins, W. S. B., and Walker, R. P. (1984), "Some Observations on Foam Sizing" Text. Chem. & Colorist, **16**, pp. 89–91.

- Qin, Z. J. (2006), Development and Application of Green Sizing Agent, *Progress in Textile Science & Technology,* **4**, pp. 5–6.

- Rust, J. P. (1992), Yarn Hairiness and the Process of Winding, *Textile Res. J.* **62**, p. 685.

- Ragheb, A., and Thalouth, I. El. (1988), "Preparation and Characterization of Carboxymethyl Starch", American Dyestuff Reporter, Vol. –77, 10, p. 41–46.

- Rajendran, B., Muthukrishnan, P., and Madhusoothanan, M. (1984), "Recent Developments in Sizing Machine and Techniques," *Ind. Text. J,* pp. 81–88.

- Garg, S. (1988), Ph. D Thesis, Dept. of Textile Tech., IIT–Delhi.

- Sejri, N., Harzallah, O., Viallier, P., Amar, S. B., and Nasrallah, S. B. (2008), Influence of Pre-wetting on the Characteristics of a Sized Yarn, *Text. Res. J* 78(4), pp. 326–335.

- Slaugon, S. D., Miller, B., and Rebenfeld, L. (1984), "Physicochemical Properties of Sized Yarns," *Text. Res. J,* **54**, p. 655.

- Saeed, I. M. (1977), The Effect of Machine Parameters on the Position of the Point of Drive in Cone Winding, Masters Dissertation, UMIST.

- Schwarz, I. G., Kovacevic, S., and Krste, D. (2011), "Analysis of Changes in Mechanical and Deformation Properties of Yarn by Sizing," *Textile Research Journal* 81, pp. 545 (originally published online on 21 October 2010).

- Schwarz, I. G., Kovacevic, S., and Dimitrovski, K. (2011), "Comparative Analysis of the Standard and Pre-wet Sizing Process" FIBERS & TEXTILES in Eastern Europe, Vol. 19, **4** (87), pp. 135–141.

- Stegmaier, T., and Trauter, J. (1990), Melliand Textilber, **71**(3), pp. 166–174.

- Schalze, U., and Trauter, J. (1993), Impact of Hairiness of Raw Yarns and Yarn Properties on Weaving Behavior of Sized Warp Yarns, *Textile Praxis Int.,* **48**(10), pp. 773–779.

- Soliman, H. A. (1995), Evaluation of Sizing as Controlling Parameter in the Tendency to Yarn Entangling, ITB Garn-und Flachenherstellung, Vol. **41**(2), pp. 42–44.

- Shruren, J. (2009), "Production and Use of Modified Starch and Starch Derivatives in China," Xuagxi Nanning Cassava Technical Development Centre, Report.

- Schwab, R. (1980), Textil. Prax. Int., **35**(3), 1980, IV

- Slauson, S. D., Miller, B., and Rebebfeld, L. (1985), *Text Res. J ,* **55**, p. 181.

- Schutz, and Enbrayat, (1989), "Foam Sizing," Melle. Text., p. 566.

- Schwab, R. (1980), Textil. Prax. Int., **35**(3), IV.

- Schoch, T. J., Radney J.A., and Maywald, E. C. (1954), Starch and Its Derivatives. *Ed.; John Wiley & Sons: New York,* pp. 637–643.

- Trommer, C. R. (1957), "Identification of Sizings on Fabrics Woven from Synthetic Filament Yarn," *Text. Res. J,* pp. 66–72.

- Trauter, J. (1984), Perspectives and Limitations of New Techniques for Applying Size to Warps, *Int. Textile Bull,* p. 27.

- Trauter, J. (1996), Progress in Sizing Research, *Int. Textile Bull,* p. 13.

- Tharanathan, R. N. (2005), Starch Value Addition by Modification, *Critical Reviews in Food Science and Nutrition,* **45**, pp. 371–384.

- Trauter, J. (1980), *Canadian Text. J* (11), pp. 62–64.

- Tsudakoma: www.tsudakoma.co.jp/eng

- Trauter, J., and Ruess, B. (1980), Mell. Text., **61**(8), p. 666.

- Trauter J. and Böttle H.(1982), Textil Prax. Int. 37 (4), 374–376 (1982).

- Vincent, J. J. et al. (1976), Text. Inst. and Ind., **14**(6), p. 191.

- Victoria, L. F., and Willett, J. L. (2005), "Reactive Extrusion of Starch-polyacrylamide Graft Coplymers: Effect of Monomer/starch Ratio and Moisture Content" Macromol. Chem. Phys. **206**, pp. 1648–1652.

- Wegener, W. (1972), Melliand Textilber, **53**(10), pp. 1161–1166.

- Wildmann, P. (1981), Textil Prax. Int., **36** (1), 23–26.

- Walkar, R. P., Perkins, W. S., and Yadon, J. L. (1980), "Foam Sizing, can Halve Energy Costs" Textile World, **130**(3), pp. 55–62.

- Wambbodri, C. G. (1986), "Foam Sizing of Cotton and Blend Yarns": Slashing trials, *Text. Res. J,* **56**(2), pp. 87–92.

- Webb, C. J., Waters, G. T., Liu, G. P., and Thomas, C. (2009), "The Influence of Yarn Count on the Splicing of Simple Continuous Filament Synthetic Yarns," *Text. Res. J,* **79**(3), 195–204.

- Wang, X. (1997), The Effect of Testing Speed on the Hairiness of Ring-spun and Sirospun Yarns, *J. Textile Inst.* **88**, p. 99.

- Xiao, H., and Zhang, W. (2009), "Current Situation of Environment Protection Sizing Agent and Paste," *J. of Sustainable Development,* Vol. **2**(3), pp. 172–175.

- Yin, X. S., and Stark, J. R. (1988), Molecular Modification of Barley Starch Granules by Different Types of Physical Treatment, *J. Cereal Sci.,* **8**, pp. 17–28.

- Zobel, H. F. (1988), Molecules to Granules: A Comprehensive Starch Review, Starch/ Stärke, **40**(2), pp. 44–50.

- Zhang, C., Yu, S. M., and Gao, X. L. (2007), Research Progress of Green Textile Size at Home and Abroad, *Progress in Textile Science & Technology,* **6**, pp. 15–16.

- Zobel, H. F. and Stephen, A. M. starch: structure, analysis and application Food Polysaccharides and their applications Ed. Stephen AM, Pholips G O and William PA, CRC Press 2006 ISBN: 978-0-9247-5922-3 Taylor and Francis Group, LLC, pp. 35–87.

5.1   Introduction

The process of passing the warp ends from weavers beam through the drop wire, heald eyes and reed dents in case of tappet and dobby weaving or through drop wire, harness cords, and dents in case of Jacquard weaving in a pre-decided order as decided by required cloth design is called drawing-in.

The ends drawn through the heald eyes or harness cords are simultaneously drawn through the dents of the reed of appropriate count and width depending upon the end density and width of the fabric (Lord and Mohamed 2005).

In Principle, drawing-in process can be divided into three parts:
1. Manual drawing-in process
2. Mechanical drawing-in process
3. Fully automatic drawing-in process, "Warpmaster"

5.1.1   Manual drawing-in process

Manual drawing-in process is performed with the help of an upright frame for mounting heald frame and the reed. It requires two persons to carry out this operation. One person finds subsequent ends from the weavers beam and gives it to the other person, who in turn draws the same through the appropriate heald eye and dent spacing (Sen Gupta 2005).

This method is more labor-oriented and very less amount of capital is required. It is the most versatile method for drawing-in any count of warp or any pattern. However, a separate drawing-in section is necessary to carry out these operations increasing the total floor space requirements. Drawing-in process is very time-consuming and the cost of drawing-in is more than any other process (Gurkan 2005).

Mechanical drawing-in process can be divided into two parts.
1. Semi-automatic reaching-in process
2. Fully automatic warp tying process

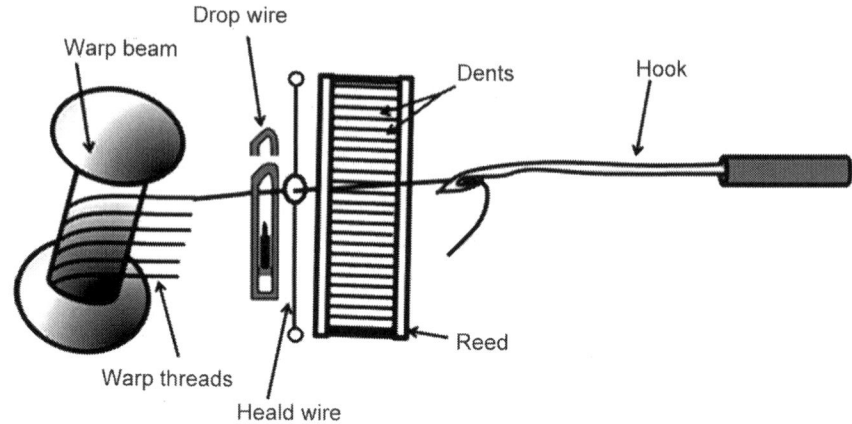

Fig. 5.1 Manual drawing-in process

Semi-automatic reaching–in process

The reaching-in machine is used for the purpose of end finding from weavers beam. In this way, this arrangement reduces the requirement of two operators to one operator per frame. In this situation, the employed operator draws the ends as provided by the reaching-in machine through appropriate heald eyes and dents in the reed according to desired fabric design (SenGupta, and Adanur 2001).

This method of drawing-in is quicker, work quality is better, and drawing-in cost is comparatively lesser than the manual drawing-in process. This process requires less flor space in comparison to manual drawing-in process. This automation requires a considerable amount of investment, hence its installation is justified only in the situation of ample work to engage the machine round the clock.

(i) Jaytex warp reaching-in machine

This semi-automatic machine is very popular in most of the big size weaving industries in India and outside world. Semi-automatic warp-reaching machine requires only one drawer. This machine is able to process woolen, cotton, synthetic, blended yarn, and paterns in warp sheet. It can be used with or without lease rod. This machine is suitable for a wide varity of dropwires, heald wires, and reed. It can be adjusted according to the beam width.

Jaytex warp reaching-in machine is available in two models; JI type and JII type. JI type warp reaching-in machine can work without lease for both sized and unsized warp beams. Either single or double yarns with both S and Z twists can be processed on this machine. The speed of selecting carriage, sorting, and progressing is controlled automatically according to the density of yarn and picking speed of the drawer. It is equipped with automatic stop motion to stop the process in case the thread is not drawn from the magazine. JII type machine is able to process leased warp yarns.

5.2   Various models of Jeytex warp tying-in machine

5.2.1   JL Model

This machine is used for tying-in leased warp threads wound on sectional warping machine or leased by machine after sizing with double lease. A wide variety of yarns like cotton, blended, woolen, P/V, filament yarns, etc., can be processed on this machine. This machine is able to handle yarns of fineness 6s Ne to 110s Ne.

5.2.2   JS Model

This machine is able to tying-in both sheets (exhausted sheet with fresh sized yarn sheet). The selection of individual warp threads takes place by selection needles. This machne is able to process 6s Ne to 80s Ne.

5.2.3   JSL Model

This machine is more versatile in comparison of previous two models. This machine is equipped with lease to lease and sheet to sheet tying-in modules. This machine can also process cotton, blended, woolen, P/V, filament yarns and spun yarns.

Fully automatic warp tying machine
This machine is able to eliminate the conventional drawing-in process completely. This machine has the facility of tying each end of running warp sheet with the corrosponding ends of the new weavers beam. The prime requirements of this process are that the new set to be woven should have the exactly same reed width and reed count, as well as the same cloth design as the previous set. Hence this type of machine is useful at the places where the same sets are run over a longer period of time. This machine requires a considerable amount of investment in comparision of both manual and semi automatic warp tying machine. This machine requires high level of attention and regular maintenance. The major advantage of this machine is its very high productivity in comparison of both manual and semi-automatic drawing-in process. The process becomes more economic if machine runs in full capacity.

This machine can work directly on loom, and eliminates the separate drawing-in section and related inventory.

Functioning of a typical automatic tying-up machine
Step 1: Fully automatic drawing-in machine separates warp yarns from each other and presents a separated end portion to a pick-up location in alignment with harness components such as a heald and/or a drop wire and/or a reed gap which themselves have been brought to their intended drawing-in positions by different, separately operated systems.

Step 2: A drawing-in member engages the yarn end and pulls it through the designed harness component. The drawing-in machine contains a drawing-in member, an arrangement for manipulating the individual harness elements, and a control stage for controlling the various functions of the drawing-in machine.

Step 3: Machine is composed of different modules to perform various functions. All modules are formed by separate units independently that are connected to one another via interfaces. All modules are controlled with the help of a high level computer of the control stage, via which the cross connections between the individual modules also run. The self dependency of modules increases their redundancy of individual functions. In this way, the dependency of individual functions on each other is negligible which simplifies the fault localization.

5.3 Yarn dressing

Yarn preparation to draw in through heald eyes, drop wires, and dents smoothly without damaging the yarn properties is called yarn dressing.

5.3.1 Jaytex yarn dressing machine

Jaytex yarn dressing machine is developed to fulfill the requirement of various height and width ranging from 120 cm to 400 cm. Jaytex frames are made of specially designed alloys that are sturdy, light weight, and have merits of high durability. Yarn clamping and tensioning devices are specially designed for efficient dressing. This machine can easily be adjusted both horizontally and vertically to adjust the height according to the loom. Clamping system of this machine is developed to get the warp yarn in parallel and firm condition. Rubber-covered tension rails are used to get firm grip to avoid the slippage of most of the slippery yarns.

In case of double beam design (for example, in terry towel weaving), the possible solutions are as under:

- One frame can cover both warp beam widths
- One single frame with double side clamping system
- Two single frame with coupling device

The double width, separable warp-tying frame is able to rationalize working on looms with double warp sheets. The separated tying frame facility of yarn dressing machine can be used for looms of normal single beam. Jaytex claims that they can manufacture the dressing frames for any kind if indigenous or imported machine.

5.3.2   Jaytex warp leasing-in machine

Leasing or lease insertion is one of the most important stages and operation of warp yarn preparation for woven fabric manufacturing. Leases are required to separate the individual threads from one another in a warp sheet to precisely define and adjust the warp thread sequence. This process through this machine establishes the warp ends in the order decided by the beamer or warper according to cloth design. Drawing-in experts can understand the importance of correct sequence in dens warp sheet of fine warp yarns like filament yarns and worsted yarns. Appropriate leasing makes the subsequent processes like drawing-in more precise, simple, and faster. The use of leasing machine improves the final quality of woven fabrics. Leasing equalizes the even tension to each warp of warp sheet. It reduces the generation of fluffs on warp yarn surface. This leasing machine is helpful to reduce end breakage rate during weaving and consequently loom stoppages are reduced. Fully automatic leasing machine does not require an operator to do this job.

Jaytex AFM2 warp leasing machine is able to make smart coordination with warp density (number of warps per centimeter in warp sheet) that becomes possible by the application of unique gear assembly with sensitive advance feelers. The warp selecting carriage moves on toothed rack that enables the selector needle to pick-up only one thread at a time. This machine is available in wide range of sizes to fulfill the requirement of weavers beam.

5.4   Fully automatic warp tying machine manufacturers

1. Shaoxing Textile Machinery Group Corp., Shaoxing, Zhejiang, China, 312000

Product: Fully automatic warp tying machine: BQ198

This machine is used to connect various warp yarns fixed on the knotting frame; it adopts world-class knotting technology, and enjoys high performance, high efficiency, high flexibility, and high quality. It is an ideal knotting solution for demands from all weaving workshops.

The machine is equipped with the latest designed knotting frame and the most reliable knotting mechanism, the preparation of warp yarn becomes simple and swift, and the knotting process is quite short, which dramatically reduces the waiting time of the weaving looms. This machine can process wool, silk, cotton, filament yarn, monofilaments, technical yarns, and other blended yarns of count range from 2 to 160 Ne. Both short fiber and long staple yarns can be processed on this machine. This machine can make maximum 600 single and double knots per minute with facility of step-less speed adjustment. Machine automatically slows down in case of 1–6 missing yarns. BQ198 is able to handle warp sheet with lease or without lease.

Staubli has developed a range of automatic drawing-in machines. DELTA and SAFIR automatic drawing-in machines are the major products of Staubli in this series. In Delta series, DELTA 100 and DELTA 110, and in Safir series, SAFIR S80 and SAFIR S30 are the major products. DELTA 100 drawing-in machine is primly used for filament weaving. It draws the yarn through heald eyes and reed dents only. DELTA 110 is improved version of DELTA 100. DELTA 100 is used where medium drawing-in requirements are registered. This machine can draw-in 140 warp threads per minute directly from the warp beam. This machine can draw-in the warp yarns into drop wires, heald eyes, and reed from one or two warp beams.

SAFIR series of drawing-in machines of Staubli is a high performance machine. SAFIR S30 drawing-in machine is used to draw-in the warp yarns to produce filament yarn fabrics and other standard textiles. This machine is able to draw the yarns through maximum of 12 healds. It consists of a mobile drawing-in unit which contains one or more than one stretching frames. This machine is able to draw the warp threads directly from weavers beam through plastic or O-shaped heald eyes and reed dents.

SAFIR S80 is comparatively a high production drawing-in machine. This machine is able to draw-in a wide range of warp yarns. This drawing-in machine is able to draw-in the warp yarns from one or two warp beams with up to two layers through drop wires, heald eyes, and reed.

5.4.1   Jaytex warp tying (beam knotting) machine

Warp tying machine can exchange the empty beam by full beam if cloth design is unchanged. Jaytex warp tying machine is made by Indigenous technology by Jayantilal S. Gandhi & Co. (Mach. Div.), Ahmedabad. This machine is compact and sophisticated and is able to tie-up very delicate yarn at a very high speed. Cotton spun yarns, filament yarns, worsted yarns, polyester-viscose (P/V) denim slub yarns, and industrial yarns can be easily processed on Taytex warp-tying machines.

Table 5.1 Comparative specifications of Staubli drawing-in machines.

Features	DELTA 100	DELTA 110	SAFIR S30	SAFIR S80
Warp widths (m)	2.3	2.3, 4.0, and 6.0	2.3	2.3/4.0
Number of warp sheets	1	1	1	1/2
Drawing-in speed ends/ min	100	140	200	160
Reed density (dents/dm)	350 (500)	350 (500)	350 (500)	350 (500)
Number of warps in 8h (ca)	Up to 5	Up to 6	Up to 10	Up to 6
Max no of heald frames (J/C healds)	20	20	–	28
Max no of heald frames (O-healds)	16	16	12	–
Drawing-in element	hook	hook	hook	hook
No of rows of drop wire path	–	1	–	1
Yarn count range	3–250	3–250	3–330	3–330

5.5 Commonly occurring defects in drawing-in process

5.5.1 Extra ends

Some warp threads are generally cast out during drawing-in process to fill for long missing ends on weavers beam.

5.5.2 Lappers

If the lappers are cut and mended after the completion of weavers beam after sizing, extra ends at drawing-in or warp tying are not required. If due to any reason, the lappers are removed no sooner these are founded, the same ends should not be selected for withdrawing or tying up.

5.5.3 Cross ends

The formation of cross ends during fabric manufacturing can be minimized by presenting the warp ends for drawing-in or warp tying properly , parallel to each other in their respective positions as in the beam. Proper combing and dressing of warp ends can minimize this problem upto some extent.

5.6   Process control in drawing-in

Drawing-in and warp tying up are the operations used to prepare weavers beam for fabric manufacturing. If weavers beam is going to prepare to run on a loom having warp stop motion facility, then warp threads have to be drawn through drop wire, heald eyes, and reed dents (Sarabhai, 1978, and Rude 2005). Two persons are required in case of conventional drawing-in process in which one is called reacher who works for selecting and presenting the ends from weavers beam, and the second person pulls the ends through drop wire, heals eyes and reed dents, called drawer. The major requirement of performing this work efficiently is given below:

The specifications of healds and reed must be check according to cloth design requirement.

- The supervisor must be aware of the concepts and principles of drawing-in process. The operator should be properly trained to do this job efficiently without making any mistake because any mistake at this stage costs a lot.
- The drawing of every end should be performed accurately to avoid any cross-ends on loom.
- Operator and supervisor should be conscious enough to suppress the incidence of extra-ends and to fill the missing warp end during fabric manufacturing on loom.
- In case of warp tying for another weavers beam, it should be ensured that the same set of heald wires and reed do not use on the loom for a long time, even if no apparent defect is noticed in them. The reason is quite clear that cleaning of healds cannot be done on the loom itself and dirty and rusty healds become the major cause of warp breakage.
- Heals and reeds should be in good condition in order to reduce warp breakage that is the major cause of fabric defect.
- The heald wires and reed from the finished beam of the loom should be cleaned properly to remove size paste, fluff, rust, etc., before using them for another new weavers beam.

5.7   References

- Gurkan, P., and Taşkin, C. (2005), "Application of Simulation Technique in Weaving Mills," Fib. and Tex. in East. Eur., 13, **3**(51), pp. 8–10.
- Lord, P. R., and Mohamed, M. H. (2005), Weaving: Conversion of Yarn to Fabric," Woodhead Publishing group Ltd.
- Gupta, R. S. (1982), Weaving Calculation, Taraporewala, Mumbai.
- Gupta, R. S. (1970), Yarn Preparation-II, Mahajan Pub. Ahmedabad.

- Adanur, S. (2001), Handbook of Weaving, CRC Press.

- Gupta, R. S., Weaving Calculation, Taraporewala , Mumbai.

- Murphy, W. S., "Handbook of weaving," Abhishek Publications, Chandigarh.

- Marks, R., and Robinson, A. T. C., "Principle of Weaving," Monara Printing Service, 91, King street, London, England.

- Singal, S. L., Technology of Synthetic Weaving, Mahajan Brothers, Ahmedabad.

- Talukdar, M. K., Sriramulu, P. K., and Ajgonkar, D. B., Weaving Mechanism: Machine and Management.

- Sarabhai, V. A. (1978), Process Control in Weaving, ATIRA, Ahmedabad.

- Arora, J. K., Modern Weaving Technology, Abhishek Publications, Chandigarh.

- Banerjee, P. K., and Alagirusamy, R. (1999), Yarn Winding, NCUTE, IIT Delhi.

- Rude, S. (2005), Weaving: The Complete Process from Thread to Cloth, available at http://www.3springshandworks.com/Documents/Process.pdf.

Index